Christ in Evolution

CHRIST
in
EVOLUTION

ILIA DELIO, O.S.F.

ORBIS BOOKS

Maryknoll, New York 10545

Founded in 1970, Orbis Books endeavors to publish works that enlighten the mind, nourish the spirit, and challenge the conscience. The publishing arm of the Maryknoll Fathers and Brothers, Orbis seeks to explore the global dimensions of the Christian faith and mission, to invite dialogue with diverse cultures and religious traditions, and to serve the cause of reconciliation and peace. The books published reflect the views of their authors and do not represent the official position of the Maryknoll Society. To learn more about Maryknoll and Orbis Books, please visit our website at www.maryknoll.org.

Copyright © 2008 by Ilia Delio.

Published by Orbis Books, Maryknoll, New York 10545-0308.

Manufactured in the United States of America.

Part of chapter 9 appeared in different form in the journal *Theology and Science*.

Queries regarding rights and permissions should be addressed to: Orbis Books, P.O. Box 308, Maryknoll, New York 10545-0308.

Library of Congress Cataloging-in-Publication Data

Delio, Ilia.
    Christ in evolution / Ilia Delio.
        p. cm.
    ISBN 978-1-57075-777-8
    1. Jesus Christ—History of doctrines. 2. Christianity—Philosophy. 3. Evolution (Biology)—Religious aspects—Christianity. 4. Evolution—Religious aspects—Christianity. 5. Religion and science. I. Title.
    BT198.D37 2008
    232—dc22
                                                              2007033112

To my colleagues at the Washington Theological Union

For their friendship and support

CONTENTS

FOREWORD

A foreword is not so much a place to summarize a book's argument as to locate or state its importance. This is indeed a very important book. Ilia Delio's spiritually enriching and theologically challenging reflections will move each reader in a unique way, but I have personally found its significance to lie especially in its timely addressing of three of the most important issues for faith and theology in an age of science.

First, Delio's work appears at a time in which the intellectual and academic worlds are becoming increasingly hospitable to scientific naturalism, the belief that nature is all there is and that scientific method is the only reliable way to understand it. This worldview has taken on notable popular form in books promoting the "new atheism" of Richard Dawkins, Sam Harris, Taner Edis, Daniel Dennett, Christopher Hitchens, and a number of others. Naturalism has been a strong temptation throughout the modern period, of course, but it has recently acquired a new surge of confidence from evolutionary biology. Today, many scientists and philosophers maintain that Darwinian ideas can explain in an exhaustively naturalistic manner all living phenomena, including human thought, morality, and spirituality.

In an indirect but significant way this learned book has much to contribute to the debate about the cognitive scope of such evolutionary naturalism. Among other achievements, Delio provides here a fresh evolutionary theology that remains completely faithful to science but without having to embrace the dead end of naturalism. She demonstrates, with great dexterity and sensitivity, how natural a setting evolutionary biology and contemporary cosmology are for retrieving and re-expressing the insights of numerous great Christian thinkers, including especially St. Bonaventure. And by taking readers through the central ideas of several, more contemporary, spiritual guides (Pierre Teilhard de Chardin, Raimon Panikkar, Thomas Merton, and Bede Griffiths), she provides a

robust alternative to the privatizing of spirituality that has weak-
ened, or in Teilhard's stronger language, "sickened" Christianity
during recent centuries. Her reflections, without being at all apologetic, provide what amounts to an invigorating and hopeful religious alternative to the inevitably pessimistic implications of
scientific naturalism on the one side and a world-escaping piety on
the other.

Second, the cosmological and evolutionary framework of this
study allows for a renewal of Christology in which the mystery of
Christ can no longer be outstripped by the temporal and spatial
magnitude of the universe. Unfortunately, while modern science
has allowed educated people to expand their thoughts and images
of the universe, Christian theology and spirituality have generally
presented the figure of Christ in dimensions too diminutive to
invite worship. A God or savior smaller than the universe will
scarcely be noticed except by those who have little interest in the
natural world. I am convinced that naturalism appeals to so many
scientifically enlightened people today at least partly because the
universe itself seems increasingly more impressive and more
resourceful than the ideas of God and Christ as these appear in
Christian education and theology. Delio's work, informed as it is by
science, provides a much needed alternative to the acosmic theology and spirituality that still shapes the religious life of countless
Christians.

Third, Delio realizes that a cosmic and evolutionary sensitivity
need not stand in contradiction to the Christian emphasis on the
personhood of God. Not everyone would agree. In my own work
I have encountered scientific thinkers who are not opposed to
"spirituality" or even some vague idea of God, but who cannot
fathom how God could be a person. The imagery suggests to them
that since persons in our experience are so much smaller than the
universe, a personal God must be also. Christian theology, on the
other hand, insists that if the universe is impersonal or less than
personal then in a sense it is ontologically smaller—less intense in
its mode of being—than human persons are. So naturalism cannot
function as an adequate spirituality either.

Resolving this difficulty is not the main focus of her work, but
Delio's cosmic Christology together with the trinitarian character
of her thought provide a good starting point for a theology that

smoothly integrates a scientific understanding of the universe with the idea of an infinite divine love and fidelity that can make sense only if God is at least personal. These and many other valuable insights await the reader of this fine book.

John F. Haught
Senior Fellow, Science & Religion
Woodstock Theological Center
Georgetown University

ACKNOWLEDGMENTS

For a number of years I have been teaching courses in Christian spirituality and I have been repeatedly impressed how the mature Christian mystic almost always arrives at a profound experience of Christ in the universe. Through the mystics I have discovered a new way of doing Christology that differs from the analytical and philosophical approach of contemporary systematic theology. This book is born out of a theological question clothed in a mystical garment, namely, what is the meaning of Christ for us today?

Our world today, marked by global consciousness and cultural and religious diversity, beckons for a new understanding of Christ. Who is the Christ for us today? How does Jesus Christ shape Christian life? Inspired by the mystics to pursue this venture, I have taken an integrative approach to plumb the meaning of Christ in an evolutionary universe. I am grateful for the sabbatical time granted me by the Washington Theological Union to write this book. The Union continues to be a school genuinely interested in the development of theology and ministry for the church today. In particular I am indebted to my colleagues at the Union who graciously read the book and offered suggestions that I believe have greatly improved it, if not saved it from error. They are John Burkhard, O.F.M., Conv., Vincent Cushing, O.F.M., and James Scullion, O.F.M. I am also grateful to Dr. Thelma Steiger and Dr. William Thompson-Uberuaga of Duquesne University who first read the manuscript, offered suggestions, and encouraged me to pursue its publication, and to John Haught, Ph.D., for his collegial support. To them and to all who have made this book possible, a heartfelt "thank you."

INTRODUCTION

Several years ago I was attending an advisory board meeting for the American Association for the Advancement of Science-Dialogue on Science, Ethics and Religion (AAAS-DoSer). We, a collection of prominent scientists and religion scholars, were sitting around the board room table, when the executive director of the National Center for Science Education, Eugenie Scott, abruptly declared, "evolution is not a theory; it is a fact!" These words struck me, as if hearing them for the first time. I had been trained as a research scientist and was familiar with the science of evolution, yet, I never really considered the difference between evolution as theory and evolution as fact. From the perspective of science, evolution is simply the way biological, chemical and physical processes proceed; indeed, much of science is concerned with mechanisms of action and the interaction of events on the physical level. Perhaps the words sounded peculiar at that moment because the debate on evolution vs. intelligent design had been in the headlines and was a "hot topic" of discussion. Ken Miller of Brown University was on the DoSer board and had just returned from Dover, Pennsylvania, where he defended the teaching of evolution in public schools. It was not long after that Cardinal Christoph Schönborn of Vienna publicly denounced evolution, especially neo-Darwinism, as an "unguided, unplanned process of random variation and natural selection," saying it was "untrue" and thus incompatible with Catholic teaching.[1] On the other hand, Father George Coyne, S.J., director of the Vatican Observatory at the time, spoke out publicly in favor of evolution, indicating that evolution is not only compatible with Catholicism but also "reveals a God who made a universe that has within it a certain dynamism and thus participates in the very creativity of God."[2]

The controversy over evolution in our time continues to rage, primarily because it upsets established religious claims. John Haught writes that "the idea of evolution is not necessarily disturbing to

1

religious people ... what is disturbing is Charles Darwin's version of evolution, which Haught summarizes in six points: "1) it offers a whole new story of creation, one that seems to conflict with the biblical accounts; 2) Darwin's notion of natural selection appears to diminish, if not eliminate, the role of God in creating the diverse forms of life; 3) Darwin's theory of human descent from 'lower' forms of life appears to question age-old beliefs in human unique-ness and ethical distinctiveness; 4) his emphasis on the prominent role of chance in evolution seems to destroy the notion of divine providence; 5) Darwinian evolution seems to rob the universe of purpose, and human life of any permanent significance; 6) and, at least for many Christians, Darwin's account of human origins seems to conflict with the notion of original sin, of the 'fall,' and therefore remove any need for a savior."[3] We might add that Darwinian evo-lution challenges the Catholic doctrine of the soul and its immedi-ate creation by God, the doctrine of creation *ex nihilo*, and the connection between sin and death.

Intelligent design is a theory that challenges the science of evo-lution, especially neo-Darwinism, by saying that life has compo-nents of irreducible complexity; the parts of physical life cannot be reduced into pieces or stages assembled over time. Rather, there is an "intelligent designer" that explains life's intricate designs.[4] The idea that an organism's complexity is evidence for the existence of a cosmic designer, however, was advanced decades before Charles Darwin was born. Its best-known exponent was English theolo-gian William Paley, creator of the famous watchmaker analogy. If we find a pocket watch in a field, Paley wrote, we immediately infer that it was produced not by natural processes acting blindly but by a designing human intellect. Likewise, he reasoned, the natural world contains abundant evidence of a supernatural creator. The argument from design, as it is known, prevailed as an explanation of the natural world until the publication of Darwin's *Origin of the Species* in 1859. The weight of the evidence that Darwin had patiently gathered swiftly convinced scientists that evolution by nat-ural selection better explained life's complexity and diversity. In some circles, however, opposition to the concept of evolution has persisted to the present. The argument from design has recently been revived by a number of academics with scientific credentials, who maintain that their version of the idea (unlike Paley's) is soundly supported by both microbiology and mathematics. These antievolutionists differ from fundamentalist creationists in that they accept that some species do change (but not much) and that the earth is much more than 6,000 years old. Like their predecessors, however, they reject the idea that evolution accounts for the array of species we see today, and they seek to have their concept—known as intelligent design—included in the science curriculum of schools.[5]

Although intelligent design holds little support among the scientific community, my intention is not to resolve this debate between evolution and intelligent design but to offer a new way of viewing these terms through a theological lens. In this book, I use the term "evolution" not only as a scientific explanation of life in the universe (a position I maintain) but in its broader meaning of dynamic change and self-transcendence in creation. One would be hard pressed not to notice an "inner pressure" within creation to move forward toward greater diversity and more complex unions. Evolution best describes this movement. I must admit, however, that I find a place for intelligent design within the context of evolution as well; that is, evolution is not entirely blind or driven by chance alone. Rather, the underlying laws of evolution that cooperate with chance reflect an overarching purpose or direction in the whole evolutionary process. I use the concept of design in this book not as a scientific account of creation but as a metaphysical term (the underlying principle of created reality) and a teleological term (the goal of the universe). The intelligent design of the universe, in my view, does not refer to the complexity of the universe but to the purpose of the universe; hence, to its "design."

To appreciate the nexus between evolution and intelligent design as overarching trends in the universe, I must explain how I arrived at the title of this book, *Christ in Evolution,* which is not meant to be provocative but to restore meaning to Christian life in a world of rapid and complex change. My primary concern in this book is not with evolution or intelligent design and their "mechanisms" of action but with the person of Jesus Christ and the meaning of Christ for Christian life today. "A Christian stands or falls with Christology," Sallie McFague writes. Being a Christian means identifying with Jesus Christ. In a world suffering from wars, famine, ecological and humanitarian crises, what does the word "Christian" mean? Or as Dietrich Bonhoeffer asked, "who really is the Christ for us today?"[6] Has the reality of Christ become irrelevant in a universe marked by evolution? Have we fallen into a type of docetism, a spiritualizing of Jesus that has led to a "Jesusolatry," to use McFague's term, an individualistic and anthropocentric type of Christian life that understands salvation purely in spiritual terms?[7] Such an understanding in her view has led to a "docile Christianity," one we might say that has become impotent to effect any real change in the world. On the other hand, we are experiencing the fallout of Western Christology that has become equally irrelevant primarily because it is an overly intellectualized tribal Christology in a world that has become global in consciousness. This intellectualized Western Christology continues to flourish in the academy, among those equipped with the latest philosophical ideas and tools of analysis. It is not theology gone awry; it is simply sterile in content and thus impotent to effect change in the Christian heart and mind, that is, the type of change that leads to a more visible expression of Christ. In my view, spirituality must be at the heart of Christology today because spirituality is of the Spirit, and where there is the Spirit, there is change. From an evolutionary perspective, the Spirit of God who hovered over this creation from the beginning continues to breathe within and among us, luring us into a new creation. Saint Paul writes that "only the Spirit can search the depths of God" (1 Cor 2:10), and thus only one filled with the Spirit can search the depths of God, especially as we consider God's incarnate presence in creation. It is because of this perspective that my interest in pursuing "Christ in Evolution" is through the insights of the mystics.

Whether we examine mystics of the early church, medieval period, or modern age, we find that virtually all mystics reveal a dynamic understanding of God and God's creation. Evolution is integral to the mystics' vision not as a science but as a means of discovering God. Bernard McGinn defines mysticism as "a direct or immediate consciousness of the presence (or absence) of God."[8] Christian mysticism rests on an intense personal relationship with Jesus Christ and conformity to Christ through a deep relationship of love. Take, for example, the profound insight of the thirteenth-century Franciscan penitent, Angela of Foligno, who described her union with Christ crucified as an exchange of love. After several years of intensely following Christ and striving to remain single-hearted in love, Angela acquired a profound spirituality that enabled her to contemplate the deeper meaning of Christ in creation. The crucified Christ, as Mary Meany points out, became Angela's dialogue partner. As she began to feel the presence of the cross within herself, her soul "liquified" in the love of God.[9] Her personal relationship with Christ became so intense that she began to experience a transformation in God. And yet, it was not simply the humanity of Christ that impelled Angela to greater union. Rather, it was precisely her experience of God in Christ that made her journey truly mystical. Her experience of Christ in his suffering humanity was, at the same time, her experience of God. Throughout her writings, she refers to Christ as the "God-man," indicating that her meditations on the sufferings of Christ were not simply a morbid attraction for pain but the experience of God's presence in Christ. This experience of the divine in the suffering humanity of Christ led Angela not only to the truth of her own humanity but to a deep understanding of God's love for the whole creation. Through her identification with the crucified Christ she discovered that the world is permeated with the goodness of God. At one point she hears God say to her, "It is true that the whole world is full of me." Then she states, "I saw that every creature was indeed full of his presence."[10] Thus, Angela came to an amazing awareness that the created world is filled with the divine presence. Nowhere is this more explicit than in one of her visions where she exclaims,

> I beheld the fullness of God in which I beheld and comprehended the whole of creation, that is, what is on this side and what is beyond the sea, the abyss, the sea itself, and everything else. And in everything that I saw, I could perceive nothing except the presence of the power of God, and in a manner totally indescribable. And my soul in an excess of wonder cried out: "This world is pregnant with God!" Wherefore I understood how small is the whole of creation— that is, what is on this side and what is beyond the sea, the abyss, the sea itself, and everything else—but the power of God fills it all to overflowing.[11]

Almost seven centuries later another mystic, who was also a scientist, Pierre Teilhard de Chardin, would proclaim that the whole evolutionary creation is a divine milieu—"pregnant with God"—

and would suggest that the cosmos is a "third nature" of Christ. Whereas Teilhard was silenced by the church and not allowed to publish any of his religious writings, Angela, an uneducated penitent, enjoyed the epithet engraved on her tombstone, "Teacher of Theologians."[12] What was declared a spiritual insight in the Middle Ages became a cause for alarm in the modern period, or, we might say, the type of mystic revered in the Middle Ages became suppressed in the modern period. However we interpret the difference between Angela and Teilhard, we would do well to plumb our Christian tradition for its wealth of spiritual insight that can inspire a way of life often caught in political battles and intellectual disputes.

Angela's insight to the cosmic Christ is consistent with Franciscan theology on the whole, which, from the roots of its founder, Francis of Assisi, emphasized the incarnation as the love of God made visible in the world. The Franciscan medieval theologian Bonaventure, a contemporary of Thomas Aquinas, did not consider the incarnation foremost as a remedy for sin but the primacy of love and the completion of creation. He recapitulated an idea present in the Greek fathers of the church, namely, Christ is the redeeming and fulfilling center of the universe. Christ does not save us *from* creation; rather, Christ is the reason *for* creation. For Bonaventure and proponents of the primacy of the Christ tradition, Christ is first in God's intention to love; love is the reason for creation. Hence, Christ is first in God's intention to create. In my view, the "intelligent design" of the universe is not outside information directing the flow of life's machinery. The design of creation is a blueprint for creation, the form or shape of its existence. In this respect the intelligent design of creation is Christ because the design of creation rests on God's freedom to love. In God's eternal wisdom, God desired to share life with another and predestined Christ from all eternity, whether or not sin ever took place. Thus, Christ is the design of the universe because Christ is first in God's intention to love. Yet, it is precisely God's freedom to love that renders creation a dynamic, unfolding process of increasing complexity. God is love and love is free; thus, God creates in freedom. Things can be themselves and follow their own inner dynamic principles. Life unfolds with a certain degree of chance and uncertainty; there is messiness in creation. The positive direction of this universe toward intelligible life, however, is not the product of a divine slot machine but the expression of a loving God. Thus, the direction of evolution is toward the maximization of goodness, especially if we maintain that the incarnation is the goal of evolution. If Jesus Christ is truly creator (as divine Word) and redeemer (as Word Incarnate) then what is created out of love is ultimately redeemed by love. The meaning of Christ is summed up in creation's potential for self-transcendent love. Bonaventure used the term "spiritual matter" to describe the orientation of matter toward spirit.[13] In his view, the whole creation is made for Christ because there is a spiritual potency within matter to receive the divine Word into it. God created matter lacking in final perfection of form, he wrote, so that by reason of its lack of form and imperfection, matter might cry out for perfection.[14] This is a very dynamic view of the material world with a spiritual potency for God, which Bonaventure saw realized in the incarnation. The idea of a spiritually potent creation means that Jesus Christ is not an intrusion into an otherwise evolutionary universe but its reason and goal.

The contemporary Franciscan theologian Zachary Hayes has found in Bonaventure's integral relationship between incarnation and creation a key to cosmic Christology in an evolutionary universe. The intrinsic connection between the mystery of creation and the mystery of incarnation means that "we discover … in Jesus the divine clue as to the structure and meaning not only of humanity but of the entire universe."[15] Rather than living with a "cosmic terror" in the face of the immensity of the universe, Hayes suggests that this evolutionary universe is meaningful and purposeful because it is grounded in Christ, the Word of God.[16] This world is not merely a plurality of unrelated things but a true unity, a *cosmos*, centered in Christ. He writes, "God created toward an end. That end as embodied in Christ points to a Christified world."[17] Bonaventure clearly viewed sin as embedded in historical reality; however, he did not limit the mystery of Christ to sin. "Christ cannot be willed by God simply because of sin."[18] The incarnation is not an afterthought on the part of God. Rather, from eternity God included the possibility of a fall of the human race and therefore structured the human person with a view to redemption. As the consummation of the created order the Incarnation is willed for its own sake and not for a lesser good such as sin.[19] It is not sin that is the cause of the incarnation but simply the excess love and mercy of God.[20] Hayes states that "in Bonaventure's view, Christ's redemptive work relates to the overcoming of sin, but it does so in a way that brings God's creative action in the world to completion. God completes what God initiates in creation and crowns it with eternal significance."[21] He notes that what may appear as a mechanical process of biological evolution (without meaning or purpose) is, on another level, a limitless mystery of productive love. "God's creative love freely calls from within the world a created love that can freely respond to God's creative call."[22] That created love is embodied in Christ in whom all of creation finds its purpose. That is why, Hayes writes, "a cosmos without Christ is a cosmos without a head … it simply does not hold together."[23] God predestined Christ not only *principaliter* but *principalius*,[24] or as Bonaventure writes, "Christ is not ordained to us but we are ordained to Christ."[25] Christ is the purpose of this universe and, as exemplar of creation, the model of what is intended for this universe, that is, union and transformation in God.

By interpreting Bonaventure's integral relationship between Christ and creation in an evolutionary context, Hayes draws several insights with regard to the nature of created reality: (1) the world at its deepest level is marked by the radical potential to receive the self-communication of the mystery of divine love into it;[26] thus, it is a world that is fit for the working out of the divine purpose;[27] (2) as the divine Word who is exemplar of creation, Christ is the model for creation so that, "what happened between God and the world in Christ points to the future of the cosmos. It is a future that involves the radical transformation of created reality through the unitive power of God's love."[28] This universe, therefore, has a destiny; the world will not be destroyed. Rather, "it will be brought to the conclusion which God intends for it from the beginning, which is anticipated in the mystery of the Incarnate Word and glorified Christ."[29]

Bonaventure's vision of a dynamic, spiritually potent material universe oriented toward the human person and spiritual union with God finds its realization in Christ. The incarnation is the center of creation in Bonaventure's view because everything that precedes this event and that follows it finds meaning in the mystery of Christ. The idea of Christ in evolution, therefore, takes as its starting point the Bonaventurian vision of the primacy of Christ. Christ is the "design" of the universe because the universe is modeled on Christ, the divine Word of God; the universe is Christologically structured. Because the universe has a "plan," we can speak of the evolution of this plan as the unfolding of Christ in the universe, who is "the mystery hidden from the beginning" (Eph 3:9).

On this note we can briefly look at the chapters of this book as they strive to explore the idea of Christ in evolution, that is, the dynamic life of Christ empowering the universe through the human person open to God, and the movement of the universe toward integration and unity in love. In chapter 1, I discuss some highlights of the new science and, in particular, the science of evolution. The purpose of this chapter is not to provide a comprehensive review of the new science but to provide a background to the paradigm shift we are experiencing today. Using the work of Karl Jaspers and Ewert Cousins, I describe this shift in terms of axial-period consciousness, which marks the emergence of human individuality, freedom, and transcendence. It is Cousins's insight to the second axial period, in particular, that provides a foundation for my thesis, namely, that spirituality is key to Christology in the second axial period.

Chapter 2 strives to provide a sweeping view of the meaning of Christ, from the Gospels through the Pauline literature up to the fathers of the church. I rely in particular on the work of the New Testament scholar N. T. Wright, who claims that Western orthodoxy has for too long had an overly lofty, detached, and oppressive view of God. It has approached Christology by assuming this view of God and has tried to fit Jesus into it. In his view, this has given rise to a "docetic" Jesus and thus an impotent view of Christ. I also discuss some of the writings of Paul (and deutero-Pauline literature) that had a profound influence on the Greek fathers such as Origen and Maximus the Confessor, whose comprehensive understanding of Christ was cosmic in scope.[30] This Greek influence filtered into the Middle Ages and helped establish the doctrine of the primacy of Christ. In chapter 3, I examine the principal reasons for the incarnation in the Middle Ages, and look at the notion of the primacy of Christ among the Franciscan theologians Alexander of Hales and Bonaventure, and Duns Scotus. I examine Bonaventure's theology, in particular, primarily because his Trinitarian theology provides a theological basis for the cosmic Christ in creation. I also discuss Scotus's doctrine of the primacy of Christ because it is clear and logically explained.

Bonaventure's integral relationship between the Trinity and creation, centered in Christ, is a profound vision of the universe—so profound that he held the meaning of Christ to be incomprehensible to the human mind. Such a mystery, he claimed, must be contemplated; spirituality is the path to theology. Bonaventure's theology is helpful to understanding Christ in the second axial period where relatedness governs the new emerging consciousness. His "method" of theology locates the context of knowledge within the context of everyday life, the fragile humanity that Christ assumed, the diversity of creatures in creation, and the dignity of the human person. His spiritual approach to theology impels us to search for truth in our world, not as ideological conviction but as that "something greater" within us that binds us together despite our differences. He offers a method of doing theology that combines experience and knowledge, spirituality and theology. His is not a scientific approach but an experiential approach, a "phenomenological" approach insofar as it is an encounter with the other, contemplating the other in wonder and awe through the power of the Spirit. In Bonaventure's view only one who is on a journey to God can really know God; faith seeks understanding through the path of love. It is in view of Bonaventure's mystical theology that I begin to examine four "guides" to the second axial period.

The first guide is the scientist-mystic Pierre Teilhard de Chardin. In chapter 4 I explore Teilhard's idea of a christic universe. As a scientist, Teilhard supported evolution as the way life proceeds in the universe, not only in terms of the physical world but in the spiritual life as well. Like Bonaventure, he held that matter is spiritually potent, and the spiritual capacity of matter is realized in the incarnation. His idea of an evolutive world is "one in which the consistence of the elements and their stability of balance lie in the direction not of matter but of spirit; in such a universe the fundamental property of the cosmic mass to concentrate upon itself, within an ever-growing consciousness, as a result of attraction or synthesis."[31] The positive and convergent movement of evolution led Teilhard to posit Christ as the inner meaning of the universe and its goal, the Omega Point. Thus he envisioned the entire evolutionary movement toward Christ through the free cooperation of human beings. He asked, "What form must our Christology take if it is to remain itself in a new world?" This too is my question as I ponder the meaning of Christ in evolution.

Our second guide to the meaning of Christ in evolution is the

Catholic Hindu scholar Raimon Panikkar. Chapter 5 discusses the work of Panikkar, who is one of the most provocative theologians today, precisely because he offers a learned theological perspective to liberate Christology from its Western, intellectual form. Through years of participating in interreligious dialogue, Panikkar has found new meaning in the person of Christ, which transcends the historical Jesus of Nazareth. Panikkar writes, "Christ is not only the name of an historical personage but a reality in our own lives."[32] He uses the term "christophany" to indicate that each person bears the mystery of Christ within. The first task of every creature, he states, is to complete and perfect his or her icon of reality.[33] Christ is the symbol of our human identity and vocation which, in its acceptance and fulfillment, is the union of all created reality in the love of God. Our participation in the mystery of Christ, therefore, lies at the basis of a healing world, a world aimed toward the fullness of the reign of God. Like Bonaventure, Panikkar sees that we must first discover the Christ within ourselves, then within others and in creation, which is the fruit of prayer, contemplation, and union with God.

The writings of these contemporary mystics, as they perceive the Christ mystery unfolding in our midst, reveal a common theme, namely, the need for prayer and contemplation—inwardness—which is explored by our last two "guides" in chapter 6. The discovery of an inner center as the renewal of living Christ today is integral to the thought of Thomas Merton and Bede Griffiths, both monks of the Benedictine tradition. Merton's insight to the "transcultural Christ," in particular, is extremely helpful in illuminating the meaning of Christ in an evolutionary world. He did not view a new emerging form of Christianity as "post Christian" religion but precisely one centered in Christ, in whom he saw the power of the risen one at work in the world. His Christology builds on finding one's identity in God, since to find God is to find oneself and to find oneself is to find Christ. Only in union with Christ, he indicated, can a person be united to the many since, as Word and center of the Trinity, Christ is both the One and the Many. William Thompson states that Merton's view of the transcultural Christ means the emergence of "a person of such inner calm and personal and cultural detachment that she is capable of recognizing and perspectivizing the genuine values present in every person and every culture."[34]

One might see a similar meaning of Christian life in the writings of Bede Griffiths, an English Benedictine monk who lived in India for many years. Bede sought to bring together Christianity and Hinduism through dialogue, prayer, and shared life. His many writings showed a Christ liberated from its Western form and, like Panikkar, included persons of other religions in the mystery of Christ. In a particular way, he sought to bridge East and West through dialogue, which became for him a participation in the Christ mystery. Scholars indicate that he followed a path of contemplation deeply influenced by Hinduism, yet without relinquishing the centrality of Christ or the essential meaning of the church. His experience as a Christian among Hindus opened up for him the depth of the mystery of Christ and the importance of interreligious dialogue as the source of fruitful growth for Christianity itself.

The value of exploring Christ in evolution through the writings of these four contemporary mystics is that through a deep inner experience of God, Christ emerges as the life-giving power of the universe. In chapter 7 I examine a "rebirth" of Christology today through the experience of the mystics. I propose that their insights to the meaning of Christ supports a new type of theology, vernacular theology, a theology borne out of experience and global consciousness, which may be more fitting to our world in evolution. The rebirth of Christology is shaped by common themes noted among the mystics, including the need for a more contemplative approach to Christology through the integration of spirituality and theology, and an emphasis on the organic nature of Christ.

In chapter 8, the insights of our contemporary writers, particularly Teilhard's notion of "co-creator," are examined in light of evolution. What all four writers concede is that human participation in the Christ mystery is at the heart of evolution, insofar as the universe is aimed toward God. Christology, therefore, is not only reflection on Christ but it is "doing Christ," allowing the mystery of Christ to unfold through the formation of our lives as christic beings in the universe. Teilhard spoke of Christianity as a "new phylum" in creation,[35] indicating that Christians are to lead the whole evolutionary movement to its fulfillment in Christ. He posited a new role for the church and envisioned a new level of evolutionary creation through a process of amorization or union through love. Christian love is to be the energy of evolution as evolution advances forward. While participation in the universe through human work and activities is the way of evolving toward the fullness of Christ, Panikkar and Griffiths perceived the unfolding of Christ through interreligious dialogue. The import of dialogue for Christ in evolution, therefore, is also discussed in this chapter.

Finally, the idea of Christ in evolution rests on the idea that we humans are in evolution. Today we see human evolution shaped primarily through technology and artificial intelligence. The rise of artificial intelligence and the development of cyborgs (human hybrids of biology and machine) is influencing the future shape of human life. In chapter 9, I explore the meaning of Christ in evolution through the influence of technology and the search for extraterrestrial life. The integration of technology in human life is of such degree that we must now consider the possibility of being techno-sapiens. We must ask, what is the meaning of incarnation in view of techno-sapiens? How is the emergence of techno-sapiens influencing the future of Christian life? The rise of technology in human life also relates to the possibility of life on other planets, insofar as advances in technology and astrobiology are raising the possibilities of extraterrestrial life. A new term, "exoChristology," has been coined by Andrew Burgess to describe the study of Christ in view of extraterrestrial life.[36] The pursuit of exoChristology allows the primacy of Christ to be raised to a new level. Since Christ is first in God's intention to love and love is the basis of creation (and all created orders), every intelligible, life-bearing world will be related to Christ and be completed by an incarnation. Christ, therefore, is the "design" of every created order, every universe, wherever life is found. Christ, the incarnate and risen Word of God, is the beginning and end of all things in God.

These nine chapters are intended to represent the unfolding of the Christ mystery, from the New Testament period where Jesus of Nazareth was recognized as the Christ to the presence of Christ on a global level, to the meaning of Christ in view of technology and extraterrestrial life. The governing hermeneutic or the lens to understand a new view of Christ today is not evolution of the physical cosmos alone but more specifically the evolution of human consciousness. We are in a new age of consciousness that signifies a new level of awareness around the globe brought about through technology and mass communication. We can no longer reflect on

Christ as if the evolution of human consciousness is an epiphenomenon, an independent or reversible phenomenon unrelated to the essence of human life or the meaning of Christ. Awareness of the global human condition, from religion to culture, language, ethnicity, geography, politics, and whatever else comprises this condition has shifted through evolutionary advance, insofar as we find this new consciousness to be one of greater complexity. From a Christian perspective we must ask, What is the meaning of Christ in light of this advance? That is the task before us, to explore the unfolding presence of the Word incarnate in a world where the fire of Christ's love has grown cold because we have failed to nurture the Christ within us. God's Word is a dynamic Word intended to empower the fullness of life through the life of the Spirit. How to rekindle this power of life, to see Christ as the integrating center of our lives and of a universe moving forward into God, is the heart of this study. What I propose here is not the end of Christology in a new millennium; rather, it is simply the beginning of a new vision and a new hope for a new creation.

CHAPTER ONE

EVOLUTION, CHRIST, AND CONSCIOUSNESS

The New Scientific Paradigm

If the book of Genesis were rewritten today, how would the story begin? In light of what the new science tells us, it might begin something like this: "In the beginning was God, filled with power and mystery, and God spoke one Word, and the Word exploded into a tiny, hot, dense ball of matter that gave rise to forces and fields, quarks and particles, all joined together like a single strand of thread." However we may understand the new science of the twenty-first century, it has certainly changed our view of the cosmos from what the original author(s) of Genesis could have known, a view that continues to unfold through the use of scientific discovery and advanced technology.

There is little doubt today from the perspective of science that evolution is the way life proceeds in the universe. From the simplest structures to complex unions, the emergent properties of life show coherence and unity as life unfolds with increased complexity. But what exactly is evolution and how does it account for life in the universe? According to the *Oxford English Dictionary* the word *evolution*, to unfold or open out, derives from the Latin *evolvere*, which applied to the "unrolling of a book."[1] The biologist Francisco Ayala writes that "contrary to popular opinion, neither the term nor the idea of biological evolution began with Charles Darwin ... It first appeared in the English language in 1647 in a nonbiological connection, and it became widely used in English for all sorts of progressions from simpler beginnings."[2] The idea of evolution emerged among the nineteenth-century biologists but was made famous by Charles Darwin in his *Origin of the Species*. What Darwin sought to show was that natural life unfolds primarily through the process of natural selection, "a process that promotes or maintains adaptation and, thus, gives the appearance of purpose or design."[3] The idea that life unfolds from more simple to complex structures now holds true not only on the level of biology but on the levels of cosmology, culture, and consciousness as well.

The area of cosmological evolution is a complex topic because of the vastness of the universe and the ongoing development of science. A brief view of the cosmos (if such a view is possible!) reveals an ancient, dynamic, and expanding universe. Scientists indicate that the universe is approximately 14.7 billion years old and began in a "Big Bang," a fiery explosion of dense, hot material.[4] Today we know that we are an intermediate-size galaxy and one of 100 billion galaxies. Our own galaxy, the Milky Way, is a mid-size galaxy consisting of 100 billion stars and stretching about 100,000 light years in diameter.[5] With Albert Einstein's theories of relativity, scientists discovered that the universe is not static but dynamic and changing. Einstein's theory of special relativity showed that time is elastic and a constituent of space. His theory of general relativity showed that gravity is not a force like other forces but is a consequence of the fact that space-time is not flat but curved or "warped" by the distribution of mass and energy in it; gravity acts to structure space.[6] The idea that space and time are dynamic quantities has changed our view of the universe from one launched from the hands of the creator in a fixed state to one that is dynamic and expanding.

In 1924 Edward Hubble, using a high-powered telescope, showed that ours was not the only galaxy in the universe, that there were many others with large empty spaces between them. Hubble concluded that the universe is not static but is expanding, since the distance between galaxies is growing all the time (a static universe would soon start to contract under the law of gravity).[7] The fact that the universe is expanding suggests that the expansion began from an extremely compressed and dense state. The discovery of background microwave radiation by A. Penzias and R. Wilson in 1965 provided support to the Big Bang theory of the universe. Using a sensitive microwave detector, they discovered extra noise outside the atmosphere, a noise that was the same day and night and throughout the year, thus beyond the solar system and beyond the galaxy, and the same in every direction. Their findings supported the work of George Gamow, who suggested that the early universe should have been very hot and dense. He predicted that radiation from the very hot early stages of the universe should still be around today.[8] Scientists now believe that the universe began at a single point (Big Bang) and rapidly inflated, and this gave rise to the forces and elements that form life in the universe.

While the new science has not entirely dispensed with the laws of physics, we know that the universe is much different from what Isaac Newton described in the late seventeenth century. The mechanistic view of the world associated with Newtonian physics has been replaced with a dynamic, open-ended view of the world in which some events are in principle unpredictable. At the infinitesimal level of the atom and its subatomic particles, quantum mechanics has uncovered a realm where time, space, and matter itself behave according to laws whose very functioning have uncertainty built into them. Quantum physics has given us a new view of matter today as not only indeterminate but also as relational. The universe seems to be inherently relational.[9] There is an interplay of chance and law in the physical fabric of the universe, according to John Polkinghorne, who notes that "both chance and necessity are indispensable partners in the fruitful history of the universe." "The role of chance does not turn evolution into a cosmic lottery," he states, "but is the way the physical world explores and realizes its potential."[10]

One of the most interesting scientific theories to emerge in the twentieth century is chaos theory. Whereas in the Newtonian world physical reality was assumed to follow rigid causal pathways, today physical reality is seen to be open and "flexible," with the capacity not only to sustain change but to sustain novelty and spontaneity. Chaos is a word that the average person hardly associates with order, and yet, the science of chaos is primarily concerned with order. The science of chaos and complexity indicates to us that nonlinear dynamical systems are characterized by spontaneous, emergent changes that give rise to new order within systems. The strange attractor is a basin of attraction within the system (but different from the system) that describes the shape of chaos or spontaneous movements of a system that deviates from the normal pattern of order. Some scientists claim that the appearance of the strange attractor means that order is inherent in chaos since the "attractor" itself is a novel pattern of order that arises spontaneously within a system. In the fluctuations of chaotic systems, randomness and unpredictability at the local level, in the presence of a strange attractor, cohere over time into new, definite, and predictable forms. As a way of explaining the unfolding of life in the universe, chaos theory underscores the nature of the universe as open to new possibilities; thus, there is a certain determinate level of order in chaos, although it is not predictable.[11]

The whole history of the universe, and particularly the history of biological life on earth, has been characterized by the steady emergence of complexity. Polkinghorne states that "the story moves from an initial cosmos that was just a ball of expanding energy to a universe of stars and galaxies; then, on at least one planet, to replicating molecules, to cellular organisms, to multicellular life, to conscious life and to humankind."[12] There is now a significant amount of evidence indicating that in these "sciences of complexity" open dynamic systems do not evolve smoothly and continuously over time in a gradual manner but do so in comparatively sudden leaps. Complex dynamic systems, whether physical, chemical, biological, or social, are regularly driven far from thermodynamic equilibrium where they reach critical bifurcation points and have a propensity to self-organize or leap abruptly into new states of increasing order and complexity.[13] Pierre Teilhard de Chardin described evolution as a "biological ascent," a movement toward more complexified life forms in which, at critical points in the evolutionary process, qualitative differences emerge. This progressive evolutionary movement, according to Teilhard, is one in which the consistence of the elements and their stability of balance lie in the direction not of matter but of spirit.[14] This movement from matter to spirit, in Teilhard's view, marked "the fundamental property of the cosmic mass to concentrate upon itself ... as a result of attraction of synthesis." Thus, he concluded, "there is only one real evolution, the evolution of convergence, because it alone is positive and creative."[15] Today, scholars speak of "mind" embedded in the physical fabric of the cosmos, an idea consistent with Teilhard's notion of convergence. Theologian Elizabeth Johnson states that consciousness is integral to the whole evolutionary process that culminates in the human spirit. She writes:

The law of complexity-consciousness reveals that ever more intricate physical combinations, as can be traced in the evolution of the brain, yield ever more powerful forms of spirit. Matter, alive with energy, evolves to spirit. While distinctive, human intelligence and creativity rise out of the very nature of the universe, which is itself intelligent and creative. In other words, human spirit is the cosmos come to consciousness.[16]

Although consciousness may be latent in the whole evolutionary process, the spiritualization of matter that leads to consciousness involves suffering and death. It is now widely acknowledged by evolutionary biologists that no significant change in creation happens without environmental stress. In light of Darwinian science we can say that there is a cruel and wasteful aspect of evolution by which living beings evolve. In the evolution of life on earth we can now identify the deep potential for life to become more complex, diverse, and elaborate following critical threshold points of environmental stress and instability. Although there are evolutionary dead ends in the history of life, and the process of evolution is punctuated with periods of crisis, instability, and catastrophe, there is also an inner logic to evolution, an inherent urge to evolve new, more complex biological forms. The process of life can transcend seemingly insurmountable obstacles and thus has a truly staggering capacity to go beyond what went before, even in the face of catastrophic mass extinctions and environmental decimations.[17] Thus we can say, according to neo-Darwinian science, evolution is a self-organizing process with an overall increase in adaptive complexity, despite periods of critical instabilities and catastrophes. The process of evolution exhibits a dynamic pattern of "crisis and renewal" across all three great domains of reality— massive explosions of supernovae in the chemical enrichment of the cosmos, mass extinctions and the relatively abrupt origination of new species in the biosphere, and socio-cultural revolutions in human history, each of which gives rise to new, more organized and more inclusive forms of life.[18] Ken Wilber maintains that any fair and balanced approach to human evolution involves what he calls the "dialectic of progress." As evolution unfolds in the world of human history, each new stage resolves certain difficulties of the previous stage but then adds new and sometimes more difficult problems of its own—problems that were not present at the preceding stage of development.[19] Whereas each new emergent stage contributes new possibilities because of a greater degree of structural organization/complexity or consciousness, there are also new potential problems. Hence, evolution is a dialectic of gains and losses, although Robert Wright describes a net gain which provides evolution with a positive direction.[20]

We find an example of this dialectic of "good news, bad news" at the very origins of human history, as portrayed in the Bible by the Judeo-Christian myth of "the fall," which some scholars suggest is a "fall upward" into more complex forms of consciousness and hence freedom.[21] The biblical account of human origins suggests that the emergence of free will in the first humans, which can be correlated with the relatively sudden growth of the complex neocortex in the fossil record, goes hand in hand with our distinctively human capacity for evil. When constructing theories of human cultural evolution, therefore, it is important to acknowledge both the significant advances and yet the profound horrors that each new stage of evolution brings forth—from hunter–gatherer tribes to mythic religious empires to global informational societies. Evolution is not a linear positive progression; rather, it is movement in tension marked by gains and losses in the development of more complex life forms. As Cameron Freeman writes, "evolution is a painful process of growth … one cannot overemphasize the growth aspect of evolution to the exclusion of the losses in the evolutionary process."[22]

The idea that evolution is necessarily a positive or beneficial affair, therefore, must be balanced by more realistic theories of human evolution, which include a dialectic of suffering as well as a dialectic of advance. For evolution in human history constantly brings forth wonderful advances, including democratic freedoms, expanding knowledge, increasing life-spans, as well as negative consequences of the way freedom is used, such as the ecological crisis, the potential for a nuclear holocaust, and global terrorism.[23] But there is a pattern that connects evolution across all domains of reality, an inherent urge for evolution to go beyond what went before it by crossing critical thresholds that generate chaos and disruption and allow for systems to reorganize themselves onto a higher, more inclusive level. We humans are part of this all-encompassing evolutionary current. We emerge from an evolutionary process and are biologically linked to the natural world, although not equal to non-human creation because of intelligence and free will. Thus, we are situated in a web of life. The same currents that run through our human blood also run through the swirling galaxies and the myriad of life-forms that pervade this planet: one and the same evolutionary current moves through all—a single self-transcending current of all-pervading energy that brings new life out of seeming catastrophe. This evolutionary current has the inherent capacity to overcome even the biggest obstacles in a sweeping advance from subatomic particles to human creativity, from hydrogen gas to the human neocortex.[24] From a theological perspective, this means that the emergence of human life is contingent on more than evolution of the physical world alone; there is an inner pressure for spiritual transcendence.

Creation and Evolution

In his book *A Window to the Divine,* Zachary Hayes writes that "a careful reading of the theological tradition prior to the modern era indicates that before the so-called Copernican revolution ... there existed a religious cosmology that involved not only the insights of faith but the physical understanding of the cosmos as it was known at that time. The breakdown of such a cosmology by the shift from a geocentric model to a heliocentric model led eventually to the isolation of theology from the development of modern science."[25] The most fundamental shift in our understanding of the cosmos is the move from the vision of a universe launched essentially in its present form by the hand of the creator at the beginning of time to a vision of the cosmos as a dynamic, unfolding chemical process, immensely large in both time and space. This shift in cosmic worldview continues to challenge the place of the human in the universe, as the human has been shifted from center stage to the growing tip of an evolutionary trend. From a Christian perspective, the unfolding dynamic universe has caused a division in our consciousness. According to Hayes, we live in two worlds. In our everyday experience we live in a culture deeply conditioned by the insights and theories of modern science. But in the context of the church, its theology and liturgy, we live in a premodern world.[26] Christian theology, he states, no longer has an effective cosmology that enables believers to relate to the world in its physical character in a way that is consistent with their religious symbols. We need to reshape our religious understanding of the world, he claims, by engaging our faith with the best insights of science concerning the nature of the physical world. In a recent lecture on science and religion, John Haught said that the very substance of Christian faith seems irreversibly intertwined with the outworn imagery of an unmoving planet nested in an unchanging cosmos. He asks, "Is it possible that the universe has outgrown the biblical God who is said to be its creator? Can Christianity and its theological interpretations find a fresh foothold in the immense and mobile universe of contemporary science, or will science itself replace our inherited spiritualities altogether, as many now see happening?"[27] Although some scientists see the process of evolution as a meaningless process suffused with blind chance, the dynamic nature of the universe speaks to us, from a Christian perspective, of the home in which a loving creator has placed us. The gift of Darwin's science to theology, Haught claims, is that it can give depth and richness to our sense of the great mystery of religion.[28] Indeed, the science of evolution can help open new windows of insight into the relationship between God and world. Evolution helps us realize that God works through the chaos of creation and is less concerned with imposing design on processes than providing nature with opportunities to participate in its own creation.

While theology depends on science for information on the concrete flow of evolutionary history, science as such can provide no framework for interpreting the ultimate levels of meaning. This is the proper task of theology.[29] Creation is not about a static world but a relationship between the dynamic being of God and a world in process of coming to be. The openness of the cosmos to what is new, its capacity to leap forward, the emergence of intelligent beings, all direct the believer to the nature of the divine presence empowering the whole cosmic process. From a theological perspective, we must admit that the one God creates the whole cosmos as one diverse but interrelated system. The same creator is present to every part of the cosmos. Creation is a relationship between God's absolute being and the finite being of creatures whereby finite beings are continuously constituted in existence by God. Creation is not something that happened at the beginning of time but is rather the continuing relationship of the world to its transcendent ground. In the history of evolution, creatures become more than they were—a process whereby reality becomes not just more than it was but essentially different. This process of transcendence reflects the spiritual nature of the physical world, the capacity of matter for spirit, and for evolution toward spiritual maturity and self-reflection. Teilhard de Chardin claimed that the whole evolutionary movement is a movement toward greater consciousness, the emergence of mind in the universe, a process he called "noogenesis."[30] In this respect we can begin to describe evolution on the level of humanity as an evolution of consciousness that expresses itself in religious belief.

Axial-Period Consciousness

One of the most helpful terms to assist in understanding the relationship between human consciousness and religious belief is "axial period." "Axial period" was used by the German philosopher Karl Jaspers in 1949 to refer to the centuries between 800 B.C. and 200 B.C. when a new kind of thinking arose in the major areas of the world: in China with Confucius and Lao-Tzu, in India with Gautama Buddha, in Persia with Zoroaster, in Greece with Thales, Pythagoras, Socrates, and Plato, and in Israel with the prophetic movement. Jaspers described this period as an axial period because "it gave birth to everything which since then, the human person has been able to be."[31] In this new age, Jaspers claimed, "man [*sic*] becomes conscious of being as a whole, of himself and his limitations. He experiences the tension of the world and his own powerlessness. He asks radical questions. Face to face with the void, he strives for liberation and redemption. By consciously recognizing his limits, he sets himself the highest goals. He experiences absoluteness in the depths of self-hood and in the lucidity of transcendence."[32] William Thompson states that "what makes this period the 'axis' of human history, even *our own* history today, is the fact that humans emerged as 'individuals' in the proper sense."[33] The significance of axial-period consciousness is apparent when considered in light of pre-axial consciousness. In the pre-axial period, the human person was cosmic, collective, tribal, mythic, and ritualistic. Myth was a way in which the human person gave meaning to his/her world through the context of stories that contained essential truths. The idea of primal consciousness as mimetic consciousness, that is a consciousness of imitation, meant that humans identified with their surroundings. Ewert Cousins notes that the pre-axial consciousness of tribal cultures was located in the cosmos and in fertility cycles of nature. Primitive persons "mimed" nature and venerated nature, which appeared as a sacred reality determining one's destiny. This created a harmony between peoples and the world of nature, a harmony expressed in myth and ritual. While primitive people were closely linked to the cosmos, they were also closely linked to one another. One gained one's identity in relation to the tribe. The strong web of interrelationships within the tribe sustained persons psychologically and energized all aspects of their lives.[34]

John Cobb has claimed that what lies at the basis of the axial period is the increasing role that rationality came to have at this time.[35] The more profound role of reason in human life had several implications. Rationality, which meant the ability to control, check, and analyze, began to supercede mythical thinking, which was governed by "projection," fantasy, and fulfillment. Axial consciousness generated a new self-awareness that included awareness of autonomy and a new sense of individuality. The human person as subject emerged. Jaspers states that, with axial consciousness, personality was revealed for the first time in history. With the emergence of the rational individual came a new sense of freedom by which the human person could make conscious and deliberate decisions.[36]

Although the world religions that emerged in the first axial period are widely divergent in their doctrines and rituals, they share a common existential thread: self-reflection and self-transcendence of the human person. Jaspers claims that "what was later called reason and personality was revealed for the first time during the axial period."[37] Cobb identifies rationality as the principal characteristic of axial-period consciousness which signals several changes: (1) the power of mythical thinking was gradually superseded and replaced by rational learning through experience; (2) a new sense of what it meant to be an individual arose, that is, a new awareness of autonomy and thus identity; (3) a new sense of freedom emerged, the sense of the "autonomous I" and thus the free, self-transcendent "I."[38] Unlike the pre-axial period, which was marked by a tribal consciousness, that is, a deep sense of the collective community or tribe, nurtured by myth and ritual, as well as a sense of relatedness to the cosmos, the first axial period was marked by individual consciousness. "Know thyself" became the watchword of Greece. The Upanishad identified the *atman,* the transcendent center of the self. The Buddha charted the way of individual enlightenment; the Jewish prophets awakened individual moral responsibility.[39] Jesus, according to Jaspers, was the last in the series of Jewish prophets and stood in conscious continuity with them.[40]

One of the most distinctive forms of spirituality that emerged in the first axial period, according to Ewert Cousins, was monasticism, the solitary search for the divine ground of being, that is, for God. Monasticism did not exist among pre-axial (primal) peoples, Cousins states, because primal consciousness did not contain the distinct center of individuality necessary to produce the monk as a religious type.[41] In the axial period, consciousness evolved from mythic awareness "governed by 'projection,' fantasy, and wish fulfillment" to critical reflection.[42] Philosophers and spiritual teachers appeared, calling the public to use the intellect to free themselves from collective consciousness, from the physical world, from myth and ritual. With the awakening of reflective subjectivity, the individual could take a stand against the collectivity, become a distinct moral and spiritual self, and embark on an individual spiritual journey.[43] Like human development itself, the axial period marked the culmination of a long process of human complexification and differentiation, an increasing expansion of "worlds" from "the immediate and mythical world of primitive man to the conventional and thus increasingly rationalized world of the great civilizations, to the post-conventional world of axial man" marked by individuation.[44] This expansion or "evolution" in human development, from myth to rationalization to individuation, is what characterizes the axial person. Although Jaspers disclaimed the notion that the axial period represents a universal stage in human evolution because it is neither irreversible nor inevitable, there is no reason not to view the emergence of this new level of consciousness as evolutionary in structure, since none of the world religions have shown a reversibility in their development.

It is within the first axial period that we can locate the original meaning of Christ and, indeed, the whole development of Christology. Eric Voegelin has identified axial consciousness in the early Israelites, and the gradual conventualizing of Israel, with the establishment of a monarchy and development of rational thinking. Israel entered into axial consciousness through its increasingly rational understanding of God, that is, neither as a tribal deity nor a national deity but as a transcendent deity, a personalized God who summoned the human to a personalized mode of existence. It is in the context of Hebraic consciousness that Jesus emerged. Thompson asks, "Are we permitted to speak of a new level of consciousness initiated by Jesus and coming into existence with the first Christian community?" Following John Cobb, he writes that "in Jesus we find the Hebraic ethically-responsible individual and the intense experience of God's immediacy, simultaneously … Jesus' sense of divine immediacy resulted in a present experience of the kingdom."[45] Thompson then states, "Jesus radicalized the Hebraic entry into axial consciousness … [his] death glaringly sums up what such a consciousness entails, and forever manifests that the style of life characteristic of radical God-centeredness is not the negation of this-worldly responsibility, but its intensification to the furthest limits. The suffering and death of Jesus were a radical manifestation that faith in God liberates the individual to accept the full implications of his freedom and responsibility."[46] The awareness of God's immediacy and a sense of ethical responsibility (e.g., feed the hungry, clothe the naked, have compassion on the poor) marked the Christian axial-period consciousness. Belief in the life, death, and resurrection of Jesus meant a breakthrough in consciousness within Jewish life and thought. "The resurrection belief completed the Hebraic entry into axial consciousness by revealing the last implications of belief in a transcendent God," a belief that life was not subject to mutability and death and decay but a higher kind of life beyond the life of nature and history.[47] Gerald O'Collins states that "in the faith of Israel belief in resurrection emerged as a corollary of theism," indicating that belief in the resurrection corresponded to an increasingly spiritual and transcendent view of selfhood.[48] Thompson claims that "Jesus' ministry and resurrection brought the process of man's spiritualization to its completion and removed every barrier to its complete emergence."[49] In some way, the empowering presence of God activated and heightened one's own self-responsibility as we see, for example, in Paul, who found himself capable of confronting every imaginable obstacle: death, life, the powers of the cosmos, present and future, and every living creature (Rom 8:35-39). This heightened sense of self-responsibility meant a heightened freedom. Through the work of the Spirit one was free from the law and could live on a new level of existence, free from the oppressive powers of this world and from the past. One could now live in a reality that radically transcended and rationalized them.[50] What emerged in Jesus was the immediacy of God's presence and, in his death and resurrection, the power of God's presence to conquer all forces threatening to destroy human individuation. Thus, belief in the resurrection enabled the emergence of the spiritual, transcendent self which knows itself to be more than natural and biological, capable of uniting with God and attaining eternal life. In this respect the church fathers declared that God became human so that we might become divine. The first axial-period consciousness of human transcendent nature fulfilled in Jesus Christ became the basis for the doctrine of Jesus Christ as true God and true human. That is, our understanding of Jesus Christ up to the present moment has been formulated on a level of human consciousness that has since moved to a new structure or level of existence—in short—that has evolved.

If we turn our attention to the twenty-first century, we can discern another transformation of consciousness, which Cousins refers to as the second axial period. Like the first, this period has been developing for several centuries and has reached a critical point. And like the first, it is effecting a radical transformation of consciousness. While the first axial period produced individual, self-reflective consciousness, the second axial period is characterized by global consciousness. The second axial period is largely the result of human creativity and inventiveness. Technology has fundamentally altered our view of the world and ourselves in the world. The tribe is no longer the local community but the global community which can now be accessed immediately via television, Internet, satellite communication, and travel. Explorations in space and satellite photographs of the earth have revealed the earth to be a luminous web of humanity and nature woven together like a quilt of many colors. The first photograph of the planet Earth in 1968, reprinted in all the major magazines, triggered immense awe as people marveled at the tiny blue marble-like globe suspended in space. From space, the earth seems like a single tribe of humanity. It is only when one begins to walk the earth that one realizes that the tribe has many different voices and religious expressions. For the first time since the appearance of human life on our planet, Cousins writes, all of the tribes, all of the nations, all of the religions are beginning to share a common history.[51] People are becoming more aware of belonging to humanity as a whole and not to a specific group. This new global consciousness must be seen from two perspectives: (1) from a horizontal perspective, cultures and religions must meet one another on the surface of the globe, entering into creative encounters that will produce a complexified collective consciousness. Cousins borrows the term "complexified collective consciousness" from Teilhard de Chardin to describe the convergence of centers of consciousness in the evolutionary process;[52] (2) From a vertical perspective, cultures and religions must plunge their roots deep into the earth in order to provide a stable and secure base for future development. This new global consciousness must be organically ecological, supported by structures that will ensure justice and peace. Cousins indicates that the second axial period is communal, global, ecological, and cosmic. It is not merely a shift from first axial period consciousness; it is an advancement in the whole evolutionary process. The second axial period challenges the religions to bring about a new integration of the spiritual and the material, of sacred energy and secular energy into a total global human energy.[53] Thus it encourages dialogue, community, and relationship with a growing awareness that each person is something of the whole. The field of quantum physics offers an understanding of the material world that radically differs from the past. Matter and energy are interrelated, and what was once understood as atoms, the building blocks of matter, are now seen to be interrelated particles. From an evolutionary viewpoint, the whole of humanity emerges from a common set of proteins and, while genetically divergent, shares the same genetic materials with lower species. The "electronic mind," the Internet, offers global connectivity and instant communication across the boundaries of languages, cultures, religions, and ethnicities. The advancement of technology and science, therefore, has rendered the second-axial-period person a global, pluralistic person, an interrelated being in search of identity and relationship. No longer is the human person content with the subjective, reflective critical awareness of the first axial period. Now one is in need of relatedness.

The Emergence of a New Christology

It is because we are aware of the interrelatedness of the earth that we can no longer maintain a homogenous, reified Western European view of Jesus Christ, a "white" Christ who is personal savior of the confessed baptized. Justo Gonzalez points out that Christology cannot be an abstract doctrine but begins with the experience of God in human history.[54] While Christology is the name for theological reflection on the mystery of Christ, it is not speculative knowledge but has a *Sitz im Leben* (a concrete or particular context). Christology is a reflection on Christ in a particular historical milieu, shaped by a particular theology, philosophy, social and political climate. In this respect, all Christology (like theology itself) is local because context, culture, language, and experience shape the understanding of Jesus Christ. The Christology of the first twenty centuries of Christian history has been forged by the unfolding of Christian belief against the background of Hebrew and Greek culture. As Teilhard wrote, "Our Christology is still expressed in exactly the same terms as those which, three centuries ago, could satisfy those whose outlook on the cosmos it is now physically impossible for us to accept.[55] Raimon Panikkar finds today's Christology narrow and stifled. He writes, "Today's Christology is not catholic or universal, nor does it need to be. Independent of its content, the very parameters of intelligibility belong to a powerful phylum of human culture, but a single phylum after all."[56] Panikkar describes Western Christology as a tribal Christology centered almost exclusively on its own concerns, with almost total neglect of other human religious experiences, a Christology for the internal purposes of Christians, perhaps even to conquer the world. Because this Christology was forged against the backdrop of Greek philosophy and Roman politics, he asks how this Christology relates to the current situation today:

> The existential situation of the world at the outset of the twenty-first century is so serious that we must not allow our-

selves to be absorbed by internal political polemics or problems of a minor order (priesthood for women, protestant
sacraments, ecumenism, sexual morality, etc.). The world is
undergoing a human and ecological crisis of planetary pro-
portions. Seventy-five percent of its population lives in subhuman conditions; thousands of children die every day
because of injustices man has perpetrated. Since 1945 wars
kill more than twelve hundred persons a day; religious intol-
erance is still all too alive throughout the planet, and the conflict among religious is still intensely inflamed.[57]

Indeed, one does wonder about the relevance of a Christology
that was developed in categories and terms of first axial-period consciousness. We too may ask, with Panikkar, what does contemporary Christology have to say about the present world in which we
live? "What bearing does the Christian response have on the problems of our day and how is all of this related to Christ?"[58]

There is no doubt that the problem of defining Jesus Christ in
fifth-century Greek philosophical terms is problematic for our current age. Panikkar points out that "nobody is saved by uttering a
simply theoretical sentence or a statement of fact."[59] Rather, the existential confession of Christ ("Jesus is Lord") is what saves. It is a personal witness of having met the reality that the name discloses. That
is why it is a "saving name," because no one can say "Jesus is Lord"
except by (or in) the Holy Spirit. This is the Christian confession:
Jesus is the Christ. Panikkar writes, "Jesus Christ as an undivided
experience is the central Christian dogma, whereby the copula *is*
drops off in order to avoid an epistemic split in the unity of Christian faith."[60] Simply put, Jesus is truly God. When Jesus asked his disciples, therefore, "Who do people say that I am?" he was asking,
"How do people experience me?" "What is their awareness of my
identity?" These are questions for us today as well. After two thousand years, our awareness of Jesus as the Christ has diminished not
because of time itself but because our awareness of the world and of
ourselves in the world has radically changed. Could it be that we no
longer have an operative Christology that speaks to our world? Are
we witnessing the elimination of Christ from the world because we
no longer grasp the significance of Christ for the world? Panikkar
reminds us that a Christology confined to doctrine alone—words
and formulas—is essentially dead. He states, "A Christology deaf to

the cries of men and especially women today would be incapable of uttering any 'word of God' whatsoever."[61] In Panikkar's view, a Christology uninvolved with the world is no Christology at all.

We are faced with an immense problem today if Christianity is to survive. We are operating out of a Christology of first axial-period consciousness, devised in an outmoded cosmology that is no longer relevant to our world. Teilhard de Chardin wrote, "the universe is not a fixed framework upon which we have simply to project the image of Christ so that we can then quietly admire it for the rest of our days."[62] If Jesus Christ is a living person and not merely a living-room statue, then we need a Christology that is organic, interrelated, and cosmic. We need a Christ who is "no longer master of the world solely because he has been proclaimed to be such but because he animates the whole universe and constitutes physically the only definitive reality in which the evolution of the world is expressed."[63] The Christ of the Gospels must be rediscovered anew if Christ is to be relevant to the world in which we live. As Teilhard wrote, "a Christ whose features do not adapt themselves to the requirements of a world that is evolutive in structure will tend more and more to be eliminated out of hand."[64] Thus we ask with Teilhard, "what form must our Christology take if it is to remain itself in a new world?" If we live in a cosmos of relatedness, not simply to one another but to the earth itself, what then is the meaning of Christ? Can it be that the Christ of the first axial period must be discovered anew?

CHAPTER TWO

A BRIEF HISTORY OF CHRIST

New Testament

The dynamic world of evolution with its interplay of chance and law is a very different world from that of Jesus of Nazareth. Not only is the whole evolutionary universe an emerging process but the concept of being within an evolutionary context has changed. The dynamism of evolution means that every entity is not a fixed isolated entity; rather, every being exists in a web of relationships that both gives rise to the entity and shapes its character.[1] What influence this new concept of reality will have on understanding the person of Jesus Christ impacts the shape of Christology in the new millennium.

According to the Judaic worldview, the universe was created by God at some definite moment in the past and ordered to a fixed set of laws. The Jews taught that the universe unfolds in a unidirectional sequence with a beginning, a middle, and an end, according to a definite historical process. The biblical world of creation described in Genesis portrays an orderly world that is hierarchical and anthropocentric. The Greek astronomer Ptolemy depicted the Earth as circular surrounded by seven spheres carrying the planets, which greatly influenced birth, death, and all phenomena in the sublunar world. As center of the cosmos, the Earth was immovable. Above the spheres of the planets there were higher spheres, the lowest of which carried the fixed stars. The whole cosmos was completely round, and the planets moved in a circular orbit. In a word, the whole cosmos had been perfectly arranged by the creator with Earth at the center of the cosmos and humans at the center of the Earth.[2] It is no mere hyperbole to say that the fixed, orderly cosmos entered into by Jesus of Nazareth was radically different from the evolutionary cosmos we know today.

My intention here is not to plumb the scholarly literature on

Jesus of Nazareth but to ask more specifically how Jesus of Nazareth was understood as the Christ. One thing we can say for certain, Christ is not the surname of Jesus, although "even in the New Testament Christ becomes a personal surname, as it is in modern usage."[3] Originally it was a title, the Greek form of "messiah." Christopher Tuckett claims that the messiahship of Jesus "only makes sense against a background of Jewish categories and beliefs. The term Messiah is really meaningless outside a Jewish context."[4] The Jews held firm to the eschatological belief in God's plan for a new age, which would be realized by a messiah.[5] Some of the language of Jewish messianic hope was applied to Jesus by early Christians. The earliest reflection on Jesus Christ was among the disciples who experienced Christ as the power of God in their midst. Jürgen Moltmann writes, "Jesus' history as the Christ does not begin with Jesus himself. It begins with the *ruach*/the Holy Spirit. It is the coming of the Spirit, the creative breath of God: in this Jesus comes forward as 'the anointed one' (*mashiah, christos*), proclaims the gospel of the kingdom with power, and convinces many with the signs of the new creation."[6] The fullness of Yahweh's *ruach/Holy Spirit*, Moltmann states, energized the works of Jesus. Where the Spirit is not active, Jesus cannot do anything either.[7] We might identify the power of the Spirit in Jesus' life as one of an intense God-relationship, one that was unconditionally intimate and intense in its quality, as Jesus' use of the expression "Abba" indicates. According to William Thompson, this God experience was undoubtedly the source of his experience of inner authority and freedom, whereby his relationships with others could transcend the legal boundaries of his time. "His proclamation of love of enemy, his association with sinners, his attempt to free the ruling class from its ideology, his association with women—all manifested the new quality of Jesus' relationship with humankind."[8]

The renowned New Testament scholar Albert Schweitzer claimed that Jesus understood himself as having a special, unique role in God's plan. He seemed to have experienced himself as the eschatological prophet, the final prophet who was anointed by God's Spirit and who was to complete the mission of the earlier prophets by announcing and enacting the good news of God's final rule.[9] Thompson describes this spiritually empowered self of Jesus as a "Christic self" marked by two dimensions: the vertical and all embracing openness to God and the horizontal openness to humanity.[10] Jesus is the one in whom the fullness of God's life sounded through. His "Abba experience" provided the source and secret of his being, message, and manner of life. Jesus shared a special intimacy with God, a special sonship; hence, he was truly a relational being, a true *person* in the sense that a God-centered otherness marked his life.[11]

Although the title of Christ (Messiah) was used by the early Christians to describe Jesus and his mission, the complexity of the title from the Old Testament could not be applied to Jesus without modification, since it easily led to popular misinterpretation. The title had to be transformed before it could be used in a Christian sense, and the agent of transformation was the life, death, and resurrection of Jesus. John McKenzie claimed that the title Messiah to describe Jesus and his mission was too narrow a description. In the New Testament it is supplemented by a number of titles that are "messianic" only in the sense that they describe some aspect of the person and mission of Jesus and define the sense in which Messiah-Christ was understood in the primitive church. According to New Testament scholars, Jesus never applied the title Messiah to himself; however, Reginald Fuller notes that Jesus' life "was implicitly Christological through and through."[12] This Christology, Fuller states, became explicit with the rise of Easter faith; that is, Jesus is manifested as messiah after his death and resurrection.[13] John McKenzie states that there are several passages in the Gospels in which a suggestion of messiahship appears to be made by Jesus. In Matt 26:63ff.; Mark 14:61ff.; Luke 22:67ff. Jesus is asked directly by the high priest if he is the Messiah. The answer varies among the Synoptic Gospels. In Mark, the answer is affirmative not of the title of Messiah, however, but of the title Son of Man. In Matthew, Jesus seems to evade a direct answer, although the title Son of Man is used. In Luke, the answer of Jesus is clearly evasive and centers around the title Son of God. In all four Gospels Jesus is asked by Pilate if he is king of the Jews, a messianic title. The answer in all three Synoptic Gospels is, "You have said it." In John 18:33, Jesus does not affirm that he is king of the Jews but that he is a king in a nonpolitical sense. Thus, it is not clear if Jesus accepted this ambiguous messianic title in this passage.[14] The question that stands out prominently in the Synoptic Gospels and which suggests Jesus' self-awareness as the Messiah is found is the question of Jesus to his disciples, "Who do you say that I am?" (Matt 16:33ff.; Mark 8:27ff.; Luke 9:18ff.). The answer is given by Peter, who identifies Jesus as the Messiah, the Christ. While the praise of Peter's faith appears only in Matthew, there is scarcely any doubt that Jesus accepts the confession. But the confession is followed immediately by a prediction of the passion, which explains the character of the messianic mission of Jesus. The theme of suffering is not drawn from any Old Testament messianic passage but from the theme of the Suffering Servant of Yahweh and from the theme of the Son of Man. Thus, the acceptance of Jesus is qualified by an interpretation that differs from the popular use of the title.[15] The Messiah is not simply the anointed one of God but the anointed one who must suffer and undergo death, which indicates the mystery of God.

In summarizing the impact of Jesus' life, N. T. Wright states that "Jesus believed he was Israel's messiah, the one through whom YHWH would restore the fortunes of his people. Jesus' radical and countercultural agenda, which subverts both the political status quo and movements of violent revolution, was focused on his awareness of his vocation. John the Baptist re-enacted the exodus in the wilderness; Jesus would do so in Jerusalem."[16] Wright goes on to say,

> Jesus' message in the Gospels constantly invokes Isaiah 40-55, in which Yahweh returns to Zion, defeats Babylon, and liberates Israel from exile. Albert Schweitzer argued a century ago that Jesus saw the Great Tribulation, the Messianic Woes, coming upon Israel and believed himself called, like the martyrs, to go ahead of Israel and take them upon himself. This would be the victory over evil; this would be the redefined messianic task. Jesus had warned that Israel's national ideology, focused then on revolutionary movements, would lead to ruthless Roman suppression; as Israel's representative he deliberately went to the place where that suppression found its symbolic focus. He drew his counter-Temple movement to a climax during Passover week, believing that as he went to his death Israel's God was doing for Israel (and hence for the world) what Israel as a whole could not do.[17]

Wright claims:

> Jesus believed something else that makes sense (albeit radical
> and shocking sense) within his own cultural, political, and the-
> ological setting. Jesus evoked, as the overtones of his own
> work, symbols that spoke of Israel's God present with God's
> people. He acted and spoke as if he were in some way a one-
> man, counter-Temple movement, as if he were gathering and
> defining Israel at this eschatological moment—the work nor-
> mally associated with Torah. He acted and spoke as the
> spokesperson of Wisdom. Temple, Torah, and Wisdom, how-
> ever, were powerful symbols of the central Jewish belief that
> the transcendent creator and covenant God would dwell
> within Israel and order Israel's life.[18]

Jesus used precisely those symbols as models for his own work. In
particular, he not only told stories whose natural meaning was that
YHWH was returning to Zion, but he acted—dramatically and symbolically—as if it were his vocation to embody that event. Wright
therefore suggests that the Temple and Yahweh's return to Zion are
the keys to the Christology of the Gospels. "Forget the titles," he
says, "at least for a moment; forget the pseudo-orthodox attempts
to make Jesus of Nazareth conscious of being the second person of
the Trinity," he states. "Forget the arid reductionism that is the mir-
ror-image of that unthinking would-be orthodoxy. Focus instead, if
you will, on a young Jewish prophet telling a story about Yahweh
returning to Zion as judge and redeemer, and then embodying it by
riding into the city in tears, by symbolizing the Temple's destruction, and by celebrating the final exodus."[19] Wright proposes, as a
matter of history, that Jesus of Nazareth was conscious of a vocation
given him by the one he knew as "Father" to enact in himself what,
in Israel's scriptures, Israel's God had promised to accomplish. He
would be the pillar of cloud for the people of the new exodus. He
would embody in himself the returning and redeeming action of
the covenant God. According to Wright, Schweitzer was right to
see that his eschatological Jesus would shake comfortable Western
orthodoxy to its foundations.[20]

To interpret the meaning of Jesus for the early Christian community was to adopt symbols and images that provided a fit for
what the early Christians experienced in Jesus and for what they

said about him. C. F. Moule insists that the New Testament did not run wild in its interpretations of Jesus, producing images of him that had little to do with what he really was or with what his followers first thought of him. Rather, there was a gradual development of new insights into the meaning of the original message and person of Jesus.[21] James D. G. Dunn summarizes the position of many New Testament scholars:

> Christology should not be narrowly confined to one particular assessment of Christ, nor should it play off one against another, nor should it insist on squeezing all the different NT conceptualizations into one particular shape, but it should recognize that from the first the significance of Christ could only be apprehended by a diversity of formulations which, though not always strictly compatible with each other, were not regarded as rendering each other invalid.[22]

Wright indicates that the story of Jesus does not generate a set of theological propositions, a "New Testament theology." It generates, as Schweitzer saw with prophetic clarity, a set of tasks. Wright claims, "The great exegetical mistake of the century (perpetrated by Schweitzer himself)—the idea that first-century Jews (including Jesus) expected the end of the world and were disappointed—has so occupied the minds of scholars that the real problem of delay has gone almost unnoticed, and people now come upon it as though it were a novelty."[23] If for Jesus, and indeed for the whole early church for which we have any real evidence, the God of Israel defeated evil once and for all on the cross, then why does evil still exist in the world? Was Jesus, after all, a failure? The New Testament answers this question with one voice. The cross and resurrection won the victory over evil, but it is the task of the Spirit, and those led by the Spirit, to implement that victory in and for the whole world. The victory is found not in the life of Jesus alone but in his death and resurrection. It is in the resurrection that the power of Jesus as the Christ is experienced.

The biblical stories of the resurrection focus on transformation and new life. Although the resurrection of Jesus impelled his followers to recognize the power of God among them in the raising of Jesus to new life, Kenan Osborne states that the major response of the disciples to the resurrection was not the experience of Jesus through eating with him or seeing him but the response of faith. "For the men and women disciples who eventually believed, this faith element was a religious experience."[24] As St. Paul wrote in his Letter to the Corinthians, "If Christ has not been raised [from the dead] then our preaching is useless and your believing it is useless" (1 Cor 15:14). Roger Haight identifies two symbols that capture the various coexistent patterns within the transformation of Jesus, resurrection and exaltation. He explains these terms as follows:

> Comparing the two symbols (resurrection and exaltation), both affirm or express that Jesus did not remain in the power of death but is alive. But they do so with different emphases. Resurrection, to be awakened, emphasizes the continuance of life; exaltation emphasizes being lifted up out of this empirical world. Resurrection tends to locate Jesus restored to life in this world where he appeared. Exaltation carries Jesus out of this world where there are no longer appearances nor a succession of events in time; Jesus' being glorified is a single mystery.[25]

When we look at the resurrection narratives in the Gospels, it is interesting that none of them actually describes the resurrection as a literal resuscitation as, for example, John describes the raising to life of Lazarus, who is immediately recognized by his family and friends. William Thompson says that "one way we can make sense of these data is to view Jesus' resurrection precisely not as a return back to life as we know it, but the entry into a qualitatively new mode of being.[26] Gerald O'Collins states that "the Resurrection was no return to earthly life and transcended any mere resuscitation of a corpse."[27] Although the resurrection took place within Second Temple Judaism, early Christians did not associate the resurrection with Jesus as Messiah but as an eschatological figure; in Jesus, death gave way to new life and the hope of new life in God.[28] The resurrection texts indicate, however, that, in the resurrection, something new happened. Jesus was changed because the disciples to whom he appeared did not recognize him immediately and not apart from tangible experience. Thompson suggests that we see in the resurrection the beginnings of a new kind of consciousness. For example, he states, the risen Jesus could only be recognized after the disciples had attained a new level of awareness. The Gospels summarize this new vision by the use of the motif of the "third day," for the "third day" was the traditional day of victory and deliverance.[29] It is essential, however, to maintain that "the one who was resurrected is no one else than Jesus, so that there is continuity and personal identity between Jesus during his lifetime and his being with God."[30] But why was Jesus not immediately recognized by his disciples after the resurrection? Why was Jesus "different"?

In his book *Jesus in the New Universe Story*, Cletus Wessels says that "the resurrection of Jesus is symbolic of a qualitatively new moment or a deeper level of human consciousness flowing from the chaos of the crucifixion." "This deeper level of consciousness," he claimed, "enabled the Jesus community to experience the meaning of death as new life for all peoples."[31] "Just as in Adam all die," Paul wrote, "so too in Christ shall all be brought to life" (1 Cor 15:22). According to Thompson, the resurrection entailed a new development in human consciousness, not only a transcendence of decay and death but a new awareness of identity, responsibility, and freedom.[32] Thomas Merton suggested that in the resurrection Jesus became the "finally integrated man," which means that "the *wholeness* of Jesus' person came to actuality through that death."[33] Thompson claims that "belief in the risen Christ ultimately seems to entail transcultural implications ... When we profess belief in a risen Christ, we are professing belief in final integration as the ultimate revelation of our faith."[34] We not only understand Christ in a new way, he suggests, but we understand ourselves in a new way by grasping the transcultural dimension of our profession of faith, that is, the dimension that overcomes separateness and division. Christ, the fully integrated person, is not *a* person but *the* Person, the integration of all human persons fully united in the one Spirit of love and thus fully integrated in relation to God. The resurrected Christ is the prolepsis of what is intended for the whole cosmos—union and transformation in God.

A Christological model in which the resurrection takes priority over the historical life of Jesus relates Christ to the cosmos not in terms of causality and historical fact but of teleology and eschatology. The Franciscan theologian Bonaventure summarized the recapitulation of the cosmos in the risen Christ when he wrote the following:

All things are said to be transformed in the transfiguration of Christ in as far as something of each creature was transfigured in Christ. For as a human being, Christ has something in common with all creatures. With the stone he shares existence; with plants he shares life; with animals he shares sensation; and with angels he shares intelligence. Thus all things are transformed in Christ since in his human nature he embraces something of every creature in himself when he is transfigured.[35]

In Bonaventure's view the spiritual potency of the whole creation is already brought to completion in the resurrection of Christ, whose glorified body becomes the perfect expression of this relationship with the cosmos. In the resurrection, Jesus is not annihilated on the cross but lives in a radically transformed mode in the presence of God for eternity. What happens in Jesus, therefore, anticipates the future of humanity and of the cosmos: not annihilation of creation but its radical transformation through the power of God's life-giving Spirit.

If Jesus is the pattern of our lives, then the power to act must come from the one who is risen and victorious in love, the Christ, who sits at the right hand of the Father and in whom the power of evil is overcome. The following of Jesus Christ cannot remain a first-century event; rather, it must be renewed in every age through a freshly drawn worldview, including new praxis, stories, symbols, and answers. Every age must discover Christ anew.

Pauline Letters

One of the earliest developments in understanding the meaning of Christ appears in the writings of Pauline and deutero-Pauline literature. These latter texts are not from Paul himself but from those who knew Paul or were affiliated with his ideas. According to Christopher Tuckett, "the question of how many of the letters in the New Testament attributed to Paul should be regarded as pseudonymous is disputed."[36] Tuckett, however, identifies six letters as pseudonymous: the three Pastoral Epistles, Ephesians, Colossians, and 2 Thessalonians. Because of their pseudonymous authorship, they can be "interpreted independently of Paul."[37]

The "captivity letters" of Paul are particularly rich in detail regarding the mystery of Christ. To this group of letters belong Philippians and Philemon. Scholars believe that these letters were written during Paul's Roman imprisonment (A.D. 61-63) and present the mature views of the Apostle who considered himself commissioned to make known the mystery of Christ (Rom 1:1-6; Eph 3:1-6).[38] In the great Christological text, Phil 2:5-11, Paul quotes an early hymn, the oldest one known from primitive Christianity.[39] The hymn praises the Christian bearer of salvation in two stanzas as a heavenly figure who humbles himself (2:6-8) and as a humbled dead person who is exalted by God above the masses (2:9-11). The hymn speaks of the deepest humiliation of this salvific figure ("he humbled himself and became obedient unto death," v. 8), and Paul adds, "even death on a cross."[40] By modifying an already existing hymn, probably sung in a service of worship, Paul links the theme of Christological self-humiliation to the church. As the church strives to live through the gospel, it is dependent on the crucified one. Becker writes, "For Christians the exaltation of Christ by God according to Phil. 2:9-11 makes Christ's self-humiliation the obligatory basic attitude toward their conduct of life. Love, to which the church of God, taught by the Spirit, is called, is comprehended as the way of humility, as is appropriate to a theology of the cross."[41] Thus, Paul links together the central Christological event of the Christian faith with Spirit and love, as they govern the church. What Paul describes is not a virtue or attitude of the earthly Jesus that he recommends for imitation. Rather, Christ is the eschatological person of salvation, the one who humbled himself and became poor for our sake, making us rich out of his poverty (2 Cor 8:9) and thus making possible for us a new way of life.

Paul's frequent use of the phrase "in Christ" in this letter affirms that Christ is a principle of grace and virtue, and these things are fulfilled in Christ: the love of God, the grace of God, spiritual enrichment, the righteousness of God which the Christian becomes, freedom, strength, faith, and love. Christ, therefore, is not an exterior principle of law or doctrine but a life in which and only in which the fulfillment of Christian grace and virtue is possible. Hence, in some instances, the phrase "in Christ" appears to designate the element or the atmosphere in which the Christian lives and acts; Christ is the instrument of the Father in creation and the preservation of the entire universe. By associating the phrase with the redemptive work of Christ, which the Father accomplished in him, Paul emphasized the saving will of God in Christ Jesus. The effect of the saving work of Christ is new life, and this new life is lived not only from Christ but in Christ. In Christ the Christian is alive to God; she or he receives eternal life in Christ Jesus our Lord. In the resurrection the dead shall be made alive in Christ. Christ is not only the agent through whom new life is conferred; he is also the sustaining cause, the principle by which the new life endures. The phrase is peculiarly apt when it is used of the church with respect to its unity in one body, a unity that suppresses all differences of race and nation (Gal 3:28). Indeed, the final unity in Christ to which God's saving acts tend is the unity of all things in Christ (Eph 1:10).

As the early church grew in its second generation, there was a development in consciousness from an understanding of Jesus' eschatological sonship to the soteriological significance of Christ in which the cosmic Christ gained prominence. First-generation believers held to a certain development in Jesus' relationship with God, especially in his resurrection, where he became the Son of God who was to come, finally, in the parousia. Jesus was the expression, therefore, even the identification of God's eternal wisdom and thus was given a cosmic, mystical presence in the unfolding of history. In the Epistle to the Colossians a vision of cosmic peace is developed that is grounded on a cosmic Christology. This deutero-Pauline letter was probably written around the time of Paul (before A.D. 60) and shows continuity with Paul's other letters but also considerable development, especially with regard to Christology.[42] The author's central theme and argument of the letter are directly related to a polemic against some Jewish Christians who held Gnostic beliefs. The principal belief was that redemption depended not on the work of Christ alone but on ruling powers who were preventing Christians from ascending to God's heaven, that is, from reaching the divine fullness.[43] Scholars agree that much of the language in this letter is borrowed from the wisdom tradition of Judaism. Moltmann notes that the description of the cosmic Christ in the Letter to the Colossians reflects what is said of wisdom in the Old Testament. He writes,

> Reconciled through his death and gathered up into his rebirth, all created being is drawn into the peace of the new

community of creation. There the self-isolation of individual creatures ends. There all acts of violence end. There ends injustice. There ends the power of death. Mutual destruction is replaced by a community of peace in which all created beings are there for one another, with one another and in one another, and through the interchange of their energies keep one another in life, for one another and together. Christ is called "the head" of this new community of creation because in him the whole fullness of the Godhead dwells bodily and through him this Shekinah overflows into the new fellowship of creation, so that with Orthodox theology we can truly talk about a "deification" of the cosmos.[44]

Thus what is said of the figure of Wisdom is applied to the person of Jesus. The claim that Christ is the "firstborn of all creation" recalls the wisdom tradition, such as Prov 8:22 ("the Lord created me in the beginning of his ways"). Tuckett states that "the precise nature of this claim is open to debate: does it imply that Christ was the first created being, or does the word *prototokos* imply rather precedence and superiority over creation?"[45] The case is not settled. What seems clear, however, is that Christ is presented as the *agent* of creation as well as the *goal* of creation. Paul himself used such an idea in 1 Cor 8:6 in claiming that Christ was the agent of creation. The writer of Colossians, however, seems to go further by claiming that somehow Christ is also the goal of creation ("all things have been created ... *for* him"). Paul used such language in some earlier writings but always of *God* as the goal of creation (see, e.g., Rom 11:36; 1 Cor 8:6). Nor is such an idea found in the wisdom literature; rather, it seems to be more of an eschatological idea anticipating the second half of the hymn, "where the role of the risen Christ is to bring/restore the created order to its intended state."[46] According to the author of Colossians, *in* Christ all perfections reside most perfectly as in their exemplary cause or model, *through* Christ in the sense that God has channeled his creative work through the meritorious activity of the incarnate Son as secondary, efficient cause, and *toward* Christ in the sense that Christ is the goal of the universe. Similarly, in 1 Cor 8:6 Paul says, "For us there is only one God, the Father, *from whom* we exist, and one Lord, Jesus Christ, *through whom* are all things and *through whom* we exist."[47]

The exaltation language continues in Col 1:17, which states that Christ is "before all things" and "in him all things hold together," but it is unclear if this refers to a temporal priority or a superiority in status. A temporal priority of wisdom can be easily paralleled (cf. Sir 1:4; Prov 8), and the idea of God's Wisdom or Logos as the bonding force that holds all things together can be found in the wisdom literature (e.g., Sir 43:26). Through Christ everything will be reconciled, "whether on earth or in heaven, making peace by the blood of his cross, through himself" (Col 1:20). The final part of this hymn claims Christ as the head of the whole body, which is glossed as "the church." According to Tuckett, "the idea of the cosmos as a body was widely used as too was the idea of God as 'head' of this body."[48] The statement therefore "underlines the extraordinarily high place given to Christ as the creator, sustainer and goal of the whole of creation."[49]

The second half of the hymn focuses on Christ as the redeemer by virtue of his resurrection (Col 1:18, "firstborn of the dead"). In Col 1:13-20, the writer emphasizes that Christ is "the firstborn of all creation, for in him were created all things in heaven and earth: everything visible and everything invisible, Thrones, Dominations, Sovereignties, Powers—all things were created through him and for him" (Col 1:15-16). By calling Jesus the "image of the unseen God" the author wants to stress the singular and unique perfection of our Lord's human nature. Christ's human nature is a representation of God's own nature in the sense that it contains every perfection that can be found in a created, human nature. The humanity of Christ shows forth or makes manifest what is hidden in the unseen nature of God: "to see me is to see the Father" (John 14:9). The phrase "firstborn of all creation" is a technical one and is found in the Hebrew law of that time, which provided that the firstborn of every family or flock was to be offered to God and redeemed (in the case of humans) or sacrificed (animals). The analogous use of the term refers to something created. Just as the firstborn of the flock or herd was of the same nature as the rest of the animals, so the "firstborn of every creature" must partake of the nature of the rest of creation, that is, must also be creature. Michael Meilach uses this phrase to argue for the primacy of Christ, saying, "this is certainly true of the Word considered in his human nature, as Incarnate; but in no way can it be true of the Word considered only in the divine nature."[50] Although Meilach's argument for the primacy of Christ is somewhat forced in light of New Testament scholarship, it is not entirely misleading. Tuckett sums up the meaning of the Colossian hymn:

> The writer of the hymn has clearly used the Jewish wisdom tradition and transferred a great deal of what is said there to the person of Jesus; but he has also expanded it considerably. Thus Jesus is now presented as the agent of creation, existing before the creation of the world and providing the goal of creation. He is too the means whereby the goal is achieved, by virtue of his death on the cross [so the probably later expansion] and supremely by his resurrection ... Some real sense of the pre-existence and pre-existent cosmic creative activity of Jesus seems to be asserted here. Jesus is here being put up into the realm of the divine in a way that exceeds much of the rest of the New Testament.[51]

Although the hymn posits an exalted role of Jesus, Christians share in that role in some way. The Christian who is "in" Christ has died and been buried with Christ in baptism and has now been raised with Christ as well (Col 2:13; cf. 3:1). As "firstborn of all creation," however, Christ enjoys a privileged place in the scheme of creation, not from the viewpoint of time but from his priority in the divine intentions. Christ, having attained eternal glory at the right hand of God through his resurrection, is the firstborn from the dead, and he communicates divine life to his members, to all humanity and angels who freely accept it. As crucified, he leads all things to their fulfillment in himself as the bond of peace and unity for every creature.

Like the Letter to the Colossians, the Letter to the Ephesians is now also considered to be a pseudonymous letter with a style different from Paul and with ideas different from his own. This letter is closest to Colossians among the Pauline corpus and in some ways complements it. Whereas Colossians focuses on the exalted nature of Christ, Ephesians centers on the church, its nature, unity, and destiny. However, this ecclesial focus builds on the exaltation of Christ found in Colossians. Ephesians takes over from Colossians the idea of the cosmic role of Christ as the head and goal of the universe. We find, for example, Eph 1:10 speaking of God's plan to "gather up all things in him (Christ), things in heaven and things on earth." Christ is made head over all things for the church, so that the mystery hidden from all eternity is not only Christ himself (Col 2:2) but now the mystery is the unity in Christ of all things (Eph 1:9-10), "or more specifically the unity of Jews and Gentiles in the single glorious church (3:4, 6)."[52] Tuckett interprets the language of pre-existence less Christologically and more ecclesiologically. He writes,

> Less certain is the reference in 1:4 which says that God "chose us in him [Christ] before the foundation of the world." ... What is pre-existent is primarily God's decision to choose "us." Yet presumably there is no question at all that we are thought of as pre-existent here. Hence it is uncertain whether the action of "choosing us in Christ" implies that Christ is thought of as pre-existent either ... The "in Christ" in verse 3 may simply mean "through" or "by means of" Christ, and ... "by means of" could mean just "on the basis of what Christ was to do." ... But the primary stress here is not really Christological as such but more "ecclesiological," that is focusing on Christians as chosen by God.[53]

In light of New Testament scholarship on Ephesians, we can agree with George Maloney that the author's description of the divine plan in this letter is an eloquent act of thanksgiving to God for all the blessings with which we have been favored in the beloved Son, Jesus Christ. The "mystery of this divine plan," to restore all things in Christ, does not mean simply reconciliation but unification of all things in Christ who is head of the body, the church. The plan of God is to create a universe in subordination to the Son, who will enter it at the proper time and, as a human being, rightfully win his place as its head, thus leading it to unity in himself. The divine plan described, therefore, in some way points to the universal primacy of Christ.[54] By comparing his sovereignty to the preeminence of the head in a human body, the author affirms that Christ is the head and goal of all creation. It can be seen that all things were created "in," "through," and "for" Christ, and all things hold together in him. He is the one who gives coherence to the universe and sustains all things. He is "the reflection of God's glory and the exact imprint of God's very being, and [thus] he sustains all things by his powerful word" (Heb 1:3). In the Letter to the Ephesians, as in Colossians, we see that Christ has universal meaning; Christ is the redeeming and fulfilling center of the universe. The whole universe is caught up in the Christ-event. When Colossians and Ephesians refer to Christ and the "powers" over which Christ rules, they are testifying to the relevance of Christ's work for everything in the created world. Thus we see in both Colossians and Ephesians that Christ the redeemer conquered the world of powers, spirits, and gods. The proclamation of "universal reconciliation" liberated believers from their fear of the world and their terror of demons. To profess Jesus as the "Christ" for the early Christians was to envision a new humanity, a new creation. Christ liberated the whole creation from sin and death. In Christ the whole cosmos is renewed.

The Patristic Fathers

The notion of cosmic Christology, rooted in scripture, attained a flowering of thought in many early Christian writers, from Irenaeus to Maximus the Confessor. George Maloney states that for many of the Greek fathers, Christ was the redeeming and fulfilling center of the universe, the one in whom the entire creation found its meaning and goal.[55] We see this idea in the profound doctrine of Irenaeus of Lyons, one of the earliest writers to extol the primacy of Christ—although his thought was not developed by the church at large. Irenaeus wrote mainly to combat error, particularly that of the Gnostics, who had set up an elaborate system of intermediaries in their doctrine of creation. Against them, he insisted on the primacy of the incarnate Word in God's creative plan. The key notions of his thought are salvation and recapitulation. Salvation is not restricted to redemption from sin; it is a more extensive process by which all are led from what he calls a state of "infancy" to that of maturity or perfection. Recapitulation is broader still; it involves nothing less than the summing up of the entire cosmos in Christ as its head. In *Adversus Haereses* he writes,

So there is, as we have shown, one Father and one Christ Jesus, who comes to us through the entire economy of salvation and sums up all things in Christ, Now in this "all," there is included also humanity who is God's handiwork. Therefore, Christ likewise sums up all in himself. The invisible becomes visible, the incomprehensible comprehensible, the impassible passible, the Word becomes human; in that way Christ sums up all this in himself, that as the Word of God holds the primacy in supercelestial, spiritual, and invisible things, so he may likewise hold the first place also in visible and bodily things, and that, taking to himself the primacy and making himself the head of the Church, he may draw all things to himself at the appointed time.[56]

Although Irenaeus's doctrine reflects a cosmic Christology, such a Christology was displaced by the more pragmatic concerns of the fathers of the church who concentrated on the practical aspects of the Christian life in its concrete historical context: sin and redemption governed their thought. The Council of Nicea (A.D. 325), for example, did not elaborate on the cosmic nature of Christ but on the relationship of the Son to the Father, in response to Arianism, which held that Christ was the firstborn creature of God and not truly divine. The result was the doctrine of *homoousios,* in which the Council of Nicea affirmed that Christ is one "in being" with the Father and hence truly divine.[57] The reason for the incarnation, according to the fathers, was "for our salvation" and hence our deification, that is, to become "like God." God became human, Athanasius proclaimed, so that we might become divine.[58]

The religious, philosophical, and political battles at the Council of Chalcedon in 451 involved concepts and terms used to identify Jesus as both divine and human. Chalcedon also attempted to resolve the problem of two "natures" fully united in one person. The Chalcedonian fathers sought to find a formula that would unite the range of opinions found in different parts of the church. John Macquarrie states that the Chalcedonian "definition must be accounted one of the most truly ecumenical documents in the history of the church."[59] R. V. Sellers writes, "In a very real sense, the Council of Chalcedon may be called the place where the three ways (Alexandrian, Antiochene and Western) met."[60] The definition at Chalcedon states,

> We all with one voice teach that it should be confessed that our Lord Jesus Christ is one and the same Son, the same perfect in Godhead, the same perfect in manhood, truly God and truly man, the Same [consisting] of a rational soul and a

body; homoousios with the father as to his Godhead, and the Same homoousios with us as to his manhood; in all things like unto us, sin only excepted ... One and the same Christ, Son, Lord, Only begotten, made known in two natures [which exist] without confusion, without change, without division, without separation; the difference of the natures having been in no wise taken away by reason of the union.[61]

While cast in the metaphysical language of its time, the intent of Chalcedon, as of Nicea, was to preserve the biblical witness that only God can be the agent of our salvation and that God's work of salvation is accomplished in and through a fully human life. Although the statement of Chalcedon may have raised more questions than it answered because of the language of the formula itself, it was meant to be a statement of faith that established a criterion for orthodoxy for all further discussion of Jesus Christ, his life, and his work. It provided a norm for Christological studies which, from this period on, would have to show how Jesus Christ is true God and true human, one and the same person, "made known in two natures (which exist) without change or confusion," that is, with no change or blending of the two natures but fully united in one person.[62] The council did not seek to render a Christology as such but a "formal outline of an adequate Christological language."[63] Although Chalcedon intended to clarify the person of Jesus as human and divine, showing how God was in Christ reconciling the world to Godself, the formula became the fixed norm of orthodoxy. All Christology after Chalcedon was to be measured according to this norm. The twentieth-century theologian Karl Rahner raised the question, Was Chalcedon the end or the beginning of Christology? Whereas the magisterium of the Catholic Church maintains the formula of Chacedon and all Christological inquiry according to this formula, the new worldview brought by the science of evolution and quantum physics compels us to revisit this formula, not to abandon it but to understand it in view of second-axial-period consciousness.

Modern Period

According to Moltmann, with the development of modern Western European theology, which looked upon the cosmic Christology of Ephesians and Colossians as mythology and speculation, cosmic Christology fell out of favor after the Middle Ages.[64] "Cosmic Christ terminology," James Lyons states, "is a product of nineteenth and early twentieth centuries. It made its appearance in Germany during the 1830s and 1840s ... and entered the English language in 1857. French theologians seem to have begun using it in 1910 though as early as 1840 there is an isolated instance in French by a non-theological writer."[65] The nineteenth-century usage of "cosmos" and its derivatives parallels usage of the Greek word *kosmos* insofar as the word refers to the world or universe as a well-ordered system or to the sum total of things. Lyons claims that cosmic-Christ language arose without close attention being paid to the usage of "cosmos" either in the Bible or in any other past work. This new terminology was intended to point beyond the narrow confines of human history to the wider creation being disclosed by science.[66] When the adjective "cosmic" is used to describe Christ, it means that Christ is the instrument in God's creative activity, the source and goal of all things, the bond and sustaining power of the whole creation; the head and ruler of the universe.[67] Basically the term relates Christ to the entire created order, emphasizing that Christ's relationship to creation extends beyond the compass of earthly humans and includes the whole cosmos.

The modern discussion of cosmic Christology was initiated by the Lutheran theologian Joseph Sittler in 1961 at the General Assembly of the World Council of Churches in New Delhi. Although Sittler was supposed to talk about the unity of the church, he spoke instead on the unity of the world, basing his discussion on the cosmic Christ hymn in Col 1:15-20. Christ is the foundation of all things, so all things have access to his cosmic redemption.[68] Paul's letter shows us that everything is claimed for God and everything is related to Christ. "From Augustine onwards," said Sittler, "Western Christendom has been marked by an inability to connect the realm of grace with that of nature. Grace has been looked upon as operative only within personal morality and history."[69] The Western church's distinction between nature and grace led to a contempt for nature that has resulted in the subjugation of nature to human domination and hence to the destruction of nature today. In his words, "the man of the Enlightenment could penetrate the realm of nature, and to all intents and purposes could take it as his sphere of sovereignty, because grace had either ignored this sector or rejected it. And with every new conquest of nature a piece of God died; the sphere of grace diminished to the degree in which structures and processes in nature were claimed by the now autonomous human being."[70] We must now take our stand, he said, not with Augustine but with Irenaeus, for whom all goodness, whether in this world or at the final consummation, manifests the grace of God.[71] Sittler went on to say, "a doctrine of redemption is meaningful only when it swings within the larger orbit of a doctrine of creation," uniting the Pauline themes of creation and redemption.[72] He further emphasized that the integration of creation and redemption is found in Christ, more specifically, the cosmic Christ: "In an age that has known the devastating effects of Hiroshima, damnation now threatens not only men and women but physical nature as well. The Church needs a life-affirming Christology of nature, a cosmic Christology."[73]

Moltmann claims that, in light of Sittler's address, we must confront the "threat to nature" with a "christology of nature" in which the power of redemption does not stop short at the hearts of men and women and their morality but extends to all of nature.[74] A new cosmic Christology, he indicates, also has implications for world religions. While Sittler's address impelled some Indian theologians present at the assembly to consider the cosmic Christ in view of other religions, Sittler himself did not address the subject of world religions. What his address did renew, however, was the idea that Christology can arrive at its completion only as a cosmic Christology. It is in light of the cosmic Christ that Moltmann writes: "All other christologies fall short and do not provide an adequate content for the experiences of the Easter witnesses with the risen Christ. If Christ is the first-born from the dead, then he cannot be merely 'the new Adam' of a new humanity. He must also be understood as the first-born of the whole creation."[75]

Although it is rather bold to view the meaning of Christ in this broad historical sweep, it is important to do so in an attempt to understand the "Christological shift" we find ourselves in as we move into the second axial period. While the New Testament experience of Jesus as the Christ became fixed in a defining Christological formula (Chalcedon), such a formulation did not diminish the fact that *faith* in Christ is the basis of new life for the world, a new life not only for the followers of Jesus but for the world

CHAPTER THREE

FRANCISCAN COSMIC CHRISTOLOGY

In a world where change is integral to the process of life, the meaning of Christ cannot remain fixed but must be relevant to the age in which Christ is proclaimed. As we enter into the second axial period, we must explore the meaning of Christ in relation to religious and cultural pluralism and to the earth itself. To plumb the meaning of Christ in the second axial period is not to abandon the past; rather, it is to revisit the riches of the Christian tradition and mine the wealth of ideas there to enlighten our search for the cosmic significance of Christ. In particular, the Franciscan view of the incarnation, which was formulated in the Middle Ages in light of Anselm's theory of satisfaction, may help shed light on the meaning of Christ today.

A brief view of the rise of Western Christology indicates a shift in emphasis from the cosmic significance of Christ as found among Eastern writers to the particular saving work of Jesus Christ described by Western writers. Following Augustine in the fourth century and Anselm of Canterbury in the eleventh century, medieval theology, led by Thomas Aquinas, affirmed that Christ came because of human sin. The emphasis on the saving work of God was expounded at great length by Augustine, who compared Jesus to a physician coming to heal a sick man. Augustine said that had there been no illness, there would have been no need to send for a physician.[1] According to Michael Meilach, however, Augustine's argument was not entirely correct. Whereas it is most certainly true to say that God became human to redeem us, it does not follow that Christ did so *only*, or even *primarily*, to redeem us. Meilach states that the simple, affirmative proposition became transformed into an exclusive one. The "illness" of humankind may be a reason for the incarnation but it is not the unique or principal

53

one. Yet, it is on the basis of this fallacy that Augustine could state that without sin (the disease) there would have been no redeemer (physician).[2]

The idea that salvation was owing to sin alone weighed on the speculations of scholastic theologians. In the eleventh century, Anselm of Canterbury formulated a "theory of satisfaction" which indicated that sin was such an affront to God's honor that divine justice demanded recompense either by satisfaction or by punishment. Anselm's combination of deep devotion and theological innovation made him the special catalyst of the distinctive Latin view of the role of the God-man. In his *Cur Deus Homo* he considered redemption as the remission of sins within the context of satisfaction. The infinite magnitude of the offense of sin requires a like satisfaction, which can be achieved only by one who is both divine (and therefore can make such satisfaction) and a human being (who is bound to make it). Following Anselm's theory of satisfaction, Western Christology focused on the sinfulness of the human person, the guilt incurred by sin, and the saving work of Christ.[3]

The problem of Christ, as patristic and medieval writers noted, did not begin on the level of history but on the level of creation itself. The Benedictine Rupert of Deutz (d. 1135), for example, objected to Anselm's position, saying that the incarnation was too magnificent a work to have been dependent on Adam's sin. In view of the many other purposes God has accomplished by becoming human, Rupert said God would still have assumed a human nature. Honorius of Autun (d. 1150) agreed with Rupert, as did the illustrious Dominican Albert the Great, although his most famous student, Thomas Aquinas, did not hold this position. Relying mainly on Augustine, Thomas said there would have been no need for an incarnation if there had been no sin to blot out. The prevailing question of the age, Would Christ have come if Adam did not sin? did not affect Thomas. He found the question so unimportant that it came to be known as *the* hypothetical question, one that only God could answer (that is, only God knows what *could have* happened.) Because human beings are of limited intelligence, he indicated, we simply do not know what might have happened if things had taken a different course.[4]

While sin was the focus of medieval Christology, there was another position that emerged from speculation on the following question: If Adam had not sinned, would the Word have assumed human nature at all? That is, would Christ have come? Although Anselm's satisfaction theory was widely accepted, it was not agreed upon unanimously, especially when the doctrine of the primacy of Christ found among the Greek fathers made its way into the writings of Western theologians.[5] This doctrine did not view the incarnation as an event due to sin; rather, it maintained that Christ is first in God's intention to love and hence to create; the principal reason for the incarnation is love, not sin. The doctrine of the primacy of Christ was less speculative than Anselm's satisfaction theory because the Greek fathers identified its roots in the New Testament, where, for example, the author of Col 1:16-17 writes, "All things were created through him and for him. Before anything was created, he existed ... and he holds all things in unity." Since the world was created for Christ (Col 1:16), it must be recapitulated or reestablished in and through him under whose power all are to be united.[6]

The Franciscan school, beginning with Alexander of Hales, looked to the Greek fathers and grounded the incarnation in the possibility of God to create and hence to become incarnate. The power to create and the power to be incarnate, in the view of the Franciscans, focused on the divine nature as such, rather than on a person of the Trinity. The question Who is Jesus Christ? thus became a *theological* question integrally related to the question What kind of God could create and become incarnate? Alexander of Hales held that one must consider the doctrine of God prior to the doctrine of the incarnation or rather that the incarnation is a central entryway to faith in a credible God. Christocentrism and theocentrism are two sides of the same coin. If the doctrine of God fundamentally relates to the question of incarnation, such a possibility could be considered only within the context of creation itself. Alexander explored the question whether God is a Trinity in God's own self, or because of a creation or an incarnation. He concluded that there is no necessity in God for either creation or incarnation. Alexander's theological foundations of Christology, therefore, began not with the person of Jesus Christ but with the question of God and the possibility of a divine nature united to a human nature. He concluded that there is no necessity in God for either creation or incarnation. Rather, the power to create and the power to be incarnate focuses on the divine nature as such rather than on a person of the Trinity.[7] Since nature refers to action, creation and incarnation find their sources in the divine nature understood as a principle of action rather than in the divine essence.[8] As Kenan Osborne states, "The Incarnation, or otherwise stated the 'insecularization of God'—God entering intimately into the created world with all its history and its materiality ... is not separate from creation. Creation and incarnation must be seen in their coterminality and interdependence ... both creation and incarnation are reflections of a credible God."[9] Similarly, Bonaventure indicated that God could not communicate himself in a finite way if he were not infinitely communicative in himself.[10] Like Alexander of Hales, Bonaventure viewed the incarnation as a central entryway to faith in a credible God, that is, a God who can communicate himself in a finite way.

Franciscan theologian Zachary Hayes notes that the understanding of the doctrine of incarnation from Alexander of Hales to Scotus, including Bonaventure, "did not limit the discussion of the meaning of Christ to the reality of the cross" but expanded it to the widest possible horizon. What these theologians did, he said, is to "perceive the possible relations between the story of Jesus and the larger picture of the world."[11] They saw that the incarnation is not an isolated event but is integral to the possibility of creation itself; one is inconceivable without the other. Because of this integral relationship between creation and incarnation, the Franciscan theologians held that "a world without Christ is an incomplete world," that is, the whole world is structured Christologically.[12] Christ is not accidental to or an intrusion in creation but the inner ground of creation and its goal. The Franciscan philosopher Duns Scotus (d. 1309) claimed that God is absolutely free and chose to create this world precisely as it is to reveal his love. He considered not what God would have done had the fall not occurred but rather what the original intent of God was relative to the incarnation. In other words, What kind of God would become incarnate?

For Scotus, the mutuality between God and human persons realized in the incarnation is grounded in the very nature of God as love. The divine initiative of love has as its primary object that creature capable of receiving the fullest measure of God's goodness and glory and who, in turn, could respond in the fullest measure. He writes, "First, God wills good for himself as the end of all things; second he wills that another be good for him. This is the moment of predestination."[13] God desires perfect order and thus intends the end (the perfection of love) rather than what is closer to the end (redemption). In Scotus's terms, God is perfect love and wills according to the perfection of that love. He did not neglect sin and the need for redemption; however, he did not view sin as the reason for the incarnation. Rather, God's love is ordered, free, and holy, and in his love he loves himself forever. God also loves himself in others, and this love is unselfish because God is the cause of all creatures. On this point Scotus states that "the predestination of anyone to glory is prior by nature to the prevision of the sin or damnation of anyone."[14] Since perfect love cannot will anything less than the perfection of love, Christ would have come in the highest glory in creation even if there were no sin and thus no need for redemption. Jesus Christ is the center and summit of all God's creative and redemptive works; all of creation is ordered to him. God, therefore, intended the highest glory as the ultimate and final end and then the incarnation as leading to that end.[15] As Allan Wolter writes, "[the primacy of Christ] makes the human nature of Christ the *motif* the Divine Architect was to carry out in the rest of creation ... after his body the visible world was sculptured. The whole universe is full of Christ."[16] Christ, therefore, is the meaning and model of creation, and every creature is made in the image of Christ. Another way of expressing this idea is that the "body" of the universe is the body of Christ. Since incarnation is the perfect mutuality between the divine and the human nature, Scotus views the summit of creation as the communion of all persons with one another and with God.

The Franciscan theologian Bonaventure, a contemporary of Thomas Aquinas, also maintained that the incarnation could not be willed because of sin, which is a lesser good. As the most noble work of God, the incarnation must be willed by God as a greater good, the noble perfection of God's love. Bonaventure, like Scotus, viewed the incarnation as the achievement and summit of creation.[17] In Jesus Christ, he indicated, the potency that lies in humanity to receive the very personal self-communication of God is realized. He grounded the incarnation in the Trinity itself, describing the Trinity as a communion of persons-in-love. The Father, he writes, who is without origin is fecund, overflowing goodness; thus, the Father is the source or fountain fullness of goodness.[18] The Son is that person eternally generated by the Father's self-diffusive goodness (*per modum naturae*); that is, the Father necessarily communicates goodness by the nature of being fountain fullness. As the total personal expression of the Father, the Son is Word, and as ultimate likeness to the Father, the Son is Image.[19] The Son/Word is generated by the Father and with the Father generates the Spirit, who is that eternal bond of love between the Father and the Son. The Spirit proceeds from the Father and the Son in an act of full freedom (*per modum voluntatis*)—his procession being the act of a clear and determinate loving volition on the part of the Father and the Son.[20]

The importance of Bonaventure's theology for the development of cosmic Christology in the second axial period is that he grounds the centrality of Christ in the nature of God as Trinity. He describes an integral relation between the Trinity, the incarnation, and the creation in that God's nature as primal, fecund mystery of self-communicative love makes possible all of God's works *ad extra*. In generating the Son, the Father speaks one Word immanent to himself in which is expressed the possibility of creation.[21] As the center of divine life, the Word is the ontological basis for all that is other than the Father, which, as an expressed Word, is vivified through the life of the Spirit. It is precisely in the relationship between the Father and the Son that one must describe creation for Bonaventure, for just as "the [divine] Word is the inner self-expression of God, the created order is the external expression of the inner Word." As the Father's self-expression, the Word is the openness of the Father to the other in all its forms. God's being as self-communicative love gives expression to its entire fruitfulness in the generation of the Son, so that in generating the Son, the Father speaks one Word immanent to himself in which is expressed the possibility of creation.[22] God's creative activity, therefore, rests in his being as triune, which is to say that God could not communicate being to the finite if he were not supremely communicative in himself.[23] Hayes writes, "if it is true that the triune God creates after his own Image, that is, after the Word, then it follows that any created reality will possess, in its inner constitution, a relation to this uncreated Word."[24]

Because the relationship between the Father and the Son is the ontological basis of all other relations, created reality bears the stamp of sonship in the deepest core of its being. As the Word is the internal self-expression of God's fecund goodness, so the world is the external objectification of that self-utterance of that which is not God. When we say that "all things are created through the Word" (John 1:3), we are saying that the Father expresses himself in the Son and this self-expression is the basis of the infinite Word of God and all finite existence as well. Creation, as a "coming into existence," emerges in this relationship between the Father and the Son. It is a finite expression of the infinite Word of God. It is caught up in the mystery of the generation of the Word from the Father and is generated out of the fecundity of God's love. Thus *all* of creation bears a relationship to the Father through the Word and the Spirit and hence through the incarnation.

While the Word expresses the creativity of the Father vivified by the Spirit, the humanity of Jesus is the fullest embodiment of that self-utterance within the created world. Although Bonaventure sees no absolute necessity for an incarnation of the Word, he does discern congruity between the mode of incarnation (the divine Word) and the mystery of creation.[25] By "congruity" Bonaventure means a "factual, positive inner relation between the inner divine reality of the Word, the extra divine reality of the Word and the reality of the Incarnation."[26] According to Bonaventure, incarnation is primarily a mystery of relation.[27] God's creative action places the created human nature of Jesus in a unique relationship to the divine. Hayes writes, "So intense is this relation that the history of Jesus of Nazareth is what the inner Word of God becomes when it is most fully spoken into that which is ontologically other than itself, that is, the human nature of Jesus."[28] The humanity of Jesus is the fullest and most perfect external Word that gives expression to the inner, eternal Word as its perfect content.[29] Thus, the Word holds a middle place between the Father and the world, and it is through the Son that the Father communicates to the world at all levels. It is precisely as Word and center that the Son is the exemplar of all creation. While at one level, the whole of the Trinity is exemplary with respect to the world, at another level the mystery of exemplarity is concentrated in a unique way in the Son, for the triune structure of God himself is expressed in him.[30] Thus, as the Word is the inner self-expression of God, the created order is the external expression of the inner Word. The created universe, therefore, possesses in its inner constitution a relation to the uncreated Word. Since the Word, in turn, is the expression of the inner trinitarian structure of God, that which is created as an expression of the Word bears the imprint of the Trinity as well.[31] The incarnation, therefore, is the perfect realization of what is potentially embedded in human nature, that is, union with the divine. In this way, Christ and the world are not accidentally related but intrinsically connected. In the incarnation, "God completes what God initiates in creation and crowns it with eternal significance."[32] The meaning of creation centered in Christ is the mystery of the Word incarnate, grounded in the mystery of divine, self-communicative love. As Hayes states, "the creative and sustaining principle of all created reality is not a mystery of arbitrariness, nor a mystery of domination and control ... it is a mystery of orderly love."[33]

The centrality of Christ as the goal and perfection of the universe renders optimism to the meaning of Christ and the universe itself and thus raises the question of sin. Does Christ atone for sin or not? Bonaventure's "congruent" relationship between incarnation and creation leads him to suggest that the primary reason for the incarnation is not sin, because he sees the incarnation as the highest work of creation and carefully avoids any necessity on the part of God. Bonaventure clearly viewed sin as embedded in historical reality; however, he did not limit the mystery of Christ to sin. "Christ cannot be willed by God *occasionaliter*, that is, simply because of sin."[34] The incarnation is not an afterthought on the part of God. Rather, from eternity, God included the possibility of a fall of the human race and therefore structured the human person with a view to redemption. As the consummation of the created order, the incarnation is willed for its own sake and not for a lesser good such as sin.[35] It is not sin that is the cause of the incarnation, but simply the excess love and mercy of God.[36] In Bonaventure's view, Christ's redemptive work relates to the overcoming of sin, but it does so in a way that brings God's creative action in the world to completion. God completes what God initiates in creation and crowns it with eternal significance.[37] His doctrine of "redemption-completion" does not negate sin; rather, sin is placed in the larger context of cosmic completion.

The beauty of Franciscan theology lies in the sacramentality of creation that emanates from God's spoken Word of love. This Word transforms what is nothing into something that lovingly reflects the heart of God. Just as the eternal, divine Word is the inner self-expression of God, the created order is the external expression of the inner Word. "God, who is the purest of love within, creates not out of any need but out of desire to manifest something of the mystery of the divine truth, goodness and beauty outwardly and to bring forth creatures capable of participating in the splendor of the divine life."[38] The inner positive relation between the world and the Word by which the world is the external expression of God means that we know the world through the Word of God, and we know the Word of God through the world. That is why Christ is the key to the truth of creation—Christ is the Word of God.

Bonaventure described the created universe as the fountain fullness of God's expressed being. As God expresses himself in creation, creation, in turn, expresses the creator. Bonaventure used two images to describe creation, mirror and book, two images that connote relationship. All of creation—rocks, trees, stars, plants, animals, and humans—in some way reflects the power, wisdom, and goodness of the Trinity. God shines through creation, and the face of God is reflected in creation precisely by the way things express themselves.[39] Creation is a mirror of God wherein the Trinity shines forth and is represented at three levels of expression: a trace (vestige), an image, and a likeness.[40] In his *Breviloquium*, Bonaventure writes,

> From all we have said, we may gather that the created world is a kind of book reflecting, representing, and describing its Maker, the Trinity, at three different levels of expression: as a vestige, as an image, and as a likeness. The aspect of vestige ("footprint") is found in every creature; the aspect of image, only in intelligent creatures or rational spirits; the aspect of likeness, only in those spirits that are God-conformed. Through these successive levels, comparable to steps, the human intellect is designed to ascend gradually to the supreme Principle, which is God.[41]

In every creature, the first person of the Trinity is reflected as the power that holds the creature in being. The second person is reflected as the Wisdom or the Exemplar by which it is created. The third person is reflected as the goodness that will bring the creature to its consummation.[42] The difference in these levels of expression reflects the degree of similarity between the creature and creator. The trace (or vestige) is the most distant reflection of God and is found in all creatures. That is, every grain of sand, every star, every earthworm reflects the Trinity as its origin, its reason for existence, and the end to which it is destined. The image, however, is only found in intellectual (human) beings. It reflects the fact that the human person is created not only according to the image of the Trinity, but as image, the human person is capable of union with the divine. Bonaventure says that those humans conformed to God by grace bear a likeness to God. In his view, every creature is understood as an aspect of God's self-expression in the world, and since every creature has its foundation in the Word, each is equally close to God (although the mode of relationship differs). God is profoundly present to all things, and God is expressed in all things, so that each creature is a symbol and a sacrament of God's presence and trinitarian life.[43] The world is created as a means of God's self-revelation so that, like a mirror or footprint, it might lead us to love and praise the creator. We are created to read the book of creation so that we may know the Author of Life. This book of creation is an expression of who God is and is meant to lead humans to what it signifies, namely, the eternal Trinity of dynamic, self-diffusive love.

The profound reflection of God throughout creation signifies that the created world is a sacrament of God. Because the world expresses the Word through whom all things are made (John 1:1), every creature is in itself a "little word" of God. In this respect, the whole creation is sacramental and incarnational; every aspect of creation is a "little incarnation" of the divine Word. Scotus used the term *haecceitas* or "thisness" to describe the unique dignity not only of human persons but of all created reality. Each created being has a distinct "thisness" that distinguishes it from other similar creatures. *Haecceitas* refers to that which is intrinsic, unique, and proper to Being itself; that which makes a singular "this" and "not-that" and which sets it off from other things like it (or of the same nature.) It can be known only by direct acquaintance, not from any consideration of more general categories. We might say it signifies that deeper reality within each being that is knowable to God alone—the sacredness of each being that cannot be duplicated or cloned.[44] Attentiveness to the other, therefore, means relating to the other not as object or idol but as icon through which the infinite goodness of God radiates. This is the basis of viewing creation as a family in which we relate to all beings as brothers and sisters. In his *Major Legend of Saint Francis*, Bonaventure showed how Francis's life deepened in Christ, and the deepening of his life in

Christ enabled him to know reality as the expressive Word of God's love. A lifetime of following the footprints of Christ led Francis to contemplate the universe as the mystery of Christ. Francis's *Canticle of Creatures*, composed at the end of his life when he was sick and blind, revealed an inner depth of union with Christ expressed in the cosmos. His prayer of praise and adoration, offered as a member of the cosmic family, together with "Brother Sun" and "Sister Moon," showed that one who lives in Christ no longer calls Christ by the name of Jesus but by the name of "brother" and "sister." To live in the experience of Christ is to live in the experience of relatedness, to be a member of the cosmic family, because Christ is the Word of God through whom all things are related.

The Franciscan doctrine of the primacy of Christ holds promise in our own time as we seek to understand the meaning of an evolutionary world and the place of humans within the context of evolution. The groundwork for the renewal of this doctrine in our age has been laid by Zachary Hayes. Hayes suggests that this evolutionary universe is meaningful and purposeful because it is grounded in Christ, the Word of God. This world is not merely a plurality of unrelated things, he states, but a true unity, a true *cosmos*, centered in Christ. While Hayes relies on Bonaventure's Christology, he also looks to Karl Rahner, who indicated that Christ is the goal toward which the whole cosmos is moving and in whom the cosmos will find its completion. Rahner viewed the movement of evolution as one from matter to spirit. Whereas spirit is natural transcendence, matter is the otherness of spirit. Matter develops out of its inner being in the direction of spirit; it is dynamic because it is always "becoming," the very act of which is self-transcendence.[45] While transcendence marks the whole creation, he states, it is with the human person that matter becomes conscious of itself. The human is the self-transcendence of living matter with a capacity for self-reflection which is spirit.[46] As the apex of the created world, the spiritualized, self-reflective human person is open to the infinite, seeking fulfillment in the absolute mystery of God,[47] that is, the open transcendence of the human person to God's absolute being constitutes the very meaning and structure of being human.

In Rahner's vision, the incarnation is a "natural fit" to the potency within creation for divine self-communication. The world at its deepest level is marked by the radical potential to receive the self-communication of the mystery of divine love into it;[48] thus, it is a world that is fit for working out the divine purpose.[49] The primacy of Christ in view of evolution means that the world is not blindly hurtling itself into an aimless expansion but is moved by Christ to Christ that God may be all in all. Hayes writes, "God creates toward an end. That end as embodied in Christ points to a Christified world."[50] The universe is not meaningless or purposeless, as some scientists say today;[51] rather, it has a divine aim that is realized in the incarnation of the Word.[52] The intrinsic relationship between Christ and creation means that "what happened between God and the world in Christ points to the future of the cosmos. It is a future that involves the radical transformation of created reality through the unitive power of God's love."[53] This universe, therefore, has a destiny; the world will not be destroyed. Rather, "it will be brought to the conclusion which God intends for it from the beginning, which is anticipated in the mystery of the Incarnate Word and glorified Christ."[54] Hayes notes that what may appear as a mechanical process of biological evolution (without meaning or purpose) is, on another level, a limitless mystery of productive love. "God's creative love freely calls from within the world a created love that can freely respond to God's creative call."[55] That created love is embodied in Christ in whom all of creation finds its purpose. In Jesus Christ, the self-communication of God's own reality, intended for everyone, takes place in an historical, tangible, and irrevocable way.[56] As God's actual, irreversible, and absolute self-giving, and the acceptance of that divine self-giving, Jesus is salvation. Only in the death of Jesus, however, is God's offer fully accepted, and in the resurrection, God's salvific promise for all people vindicated.[57] Christ is the purpose of this universe and the model of what is intended for this universe, that is, union and transformation in God.

The late Scottish philosopher John Macmurray said that "the universe is God's one action informed by one intention and that God's intention, in creating the universe, is to create a place where human beings can enjoy community with the Trinity and thus with one another."[58] Franciscan theology helps us appreciate that Trinity means God is relational, self-communicative, and personal love. God is a communion of persons in love. Because God is relational, relationship is at the heart of Christ who, as divine Word, is the center of the Trinity and hence the center of creation. Christ is the truly human one who is fully open to God as love and in whom

God's loving plan for creation is revealed, namely, the unity of all things in love. Christ is truly the sacrament of God's love in the universe and, we might say, its divine aim. In the words of Zachary Hayes, "a cosmos without Christ is a cosmos without a head ... it simply does not hold together. But with Christ, all the lines of energy are coordinated and unified ... all is finally brought to its destiny in God."[59] The whole universe is designed and created with a view toward Christ, who is the noble perfection of the universe, its goal and center.[60] Stated in more contemporary terms, Christ is the "more" of an evolutionary universe. That is, Christ is both the revelation of God as love and the future of the universe in view of this love. If God's single aim in creating the universe is a communion of all living beings in love, then there is no other reason for Christ except the reason of love. The meaning of Christ therefore extends to all people, the Earth, all planets, indeed, to the entire universe and all universes.

As we enter into the second axial period we need to rediscover Christ not only as healer of wounded humanity and of the earth itself, but Christ as the meaning and goal of this universe, the sacrament of unity in love, the one who is the integrating center of the cosmos. To make this discovery we need to see the mystery of Christ with new eyes and listen to the voice of Christ in new languages. We need to set aside our preconceived notions of Christ and allow ourselves to experience the mystery of God's love anew. We need a conversion to Christ. How will this conversion come about? In the next few chapters we will explore the mystery of Christ through four contemporary spiritual writers who open up to us a new way of doing Christology in the second axial period. It is a way through prayer and contemplation, the way of the mystic.

CHAPTER FOUR

TEILHARD DE CHARDIN: THE CHRISTIC UNIVERSE

Doing Christology in an evolutionary world means that change must be at the heart of understanding Christ and Christ must be at the heart of change. It is not simply to be aware of religious, cultural, and ethnic differences; rather, it is to be aware that differences are essential to the meaning of Christ. To define Christ in terms of change and difference is a radical shift from the Greek philosophical understanding of matter and being that governed the development of Christology in the patristic period and the Middle Ages. Then, it was sufficient to arrive at an understanding of matter as substance and accident. Today, the whole understanding of matter is under revision, and we must see how any nature comprised of matter and which constitutes being arises from a complexity of relationships in an evolving web of life.

It is difficult to arrive at a new understanding of Christ, however, with our current intellectual tools. Since the time of Descartes' *Discourse on Method*, Western thought has been preoccupied with methods of analyses that separate out reason from the "stuff" of life. Contemporary Christology, at least since the nineteenth century, has been governed by historical-critical analysis. Although this method, which focuses on the human life of Jesus, is meant to correlate, interpretatively, with contemporary human life, such analysis often obscures the meaning of Christ as the power of God to act. One comes to know a lot *about* Jesus without necessarily *knowing* Jesus, for such knowledge cannot be analytically defined but is experiential. I do not intend here to dismiss historical-critical analysis since intellectual integrity leads to deepened understanding without succumbing to the purely private devotional. Indeed, both intellectual study and experience are ideally complementary; yet, one cannot help notice the priority of the intellectual over the experiential from the High Middle Ages into the modern period. However we conceive of Christology, theory and practice cannot be separate, since "knowledge of Christ in his meaning for us is itself a praxis, and we have to answer for it as such."[1]

The problem with contemporary Christology today lies in its Western form, which is intellectual, abstract, and logical. Since the great ecumenical councils of the first six centuries, which provided the architectonics of Christology, not much as changed. While the modern intellectual tools of historical-critical analysis and hermeneutical theory have given us a more accurate portrait of Jesus in his historical context, such knowledge does not penetrate the meaning of Christ. We probably know more today about Jesus of Nazareth than ever before but do we know Jesus as the Christ? Moreover, the fruits of modern Christology have been largely academic. On the parochial level we find a rudimentary understanding of Jesus Christ that is devotional and pietistic. The intellectual study of Christ in the modern/postmodern period has not impacted Christian spirituality to any significant degree beyond medieval devotion. The whole meaning of Christ has become so obscure that if the title "Christ" was dropped from the name of Jesus, it would probably make little difference in the life of the Christian faithful. We can continue the discussions on Christ in the academy, but unless insights gleaned from these discussions impact the wider church (and thus Christian life in the world) they are, in the end, of little importance. Moltmann sums up the problem of contemporary Christology well when he writes,

> We have to look at Jesus' humanity in order to know his divinity, and we have to contemplate his divinity so as to know his humanity. Anyone who resolves this dialectical process of perception into dogmatic alternatives, is resolving—or dissolving—christology altogether. That person will end up with a theological christology without Jesus, or with an anthropological Jesuology without God.[2]

Moltmann argues that we must find another way "to turn to the more complex process of coming to know Jesus Christ."[3] This new way, I believe, must reflect the shift in consciousness from first-axial-period to second-axial-period consciousness. What makes Rahner's work attractive is its existential and mystical appeal; that is, "the openness to God as question that characterizes the human spirit, and supernatural experience, God's coming to meet and answer humanity in the Incarnate Word."[4] His doctrine of the supernatural existential and the preeminence of grace enables Rahner to break through the doors of neoscholastic theology and engage a transcendental theology that is thoroughly mystical in content. The paradigm shift inherent in Rahner's theology reflects his attempt to address the human longing for fulfillment, especially in view of religious, cultural, and scientific changes. The success of his theology in the twentieth century leads us to suggest that retrieving the mystical element in Christianity may be essential to developing theology in the second axial period of the third millennium. In the next few chapters we will look at several spiritual writers who have developed a spiritual/mystical approach to doing Christology in our time. While their insights are filled with theological content, they arise from a deeper center within them, a contemplative inwardness that reflects a profound experience of Christ. Their mystical approach to Christology combines experience and knowledge, spirituality and theology, searching the depths of the Christ mystery unfolding in our midst.

The first of these theologians who received little recognition during his lifetime and suffered the censorship of his theological writings was the scientist and mystic Pierre Teilhard de Chardin (1881-1955). Born into a Catholic family in Sarcenat, France, Teilhard entered the Society of Jesus at the age of eighteen and earned a doctorate in paleontology from the Sorbonne at twenty-two. He taught at the Jesuit college in Cairo and, except for service as a stretcher bearer in World War I, spent most of his life in geological and paleontological research around the world. He was steeped in Ignatian spirituality through his formation as a Jesuit; yet as a scientist he was led to a penetrating insight of Christ in the material universe. He began writing on Christ and evolution in the 1920s and continued up to his death in 1955. He lived through two world wars and witnessed the modernist crisis, enjoying a correspondence with the modernist scholar Maurice Blondel. Teilhard discovered the writings of Scotus rather late in life, through the Sicilian Franciscan Father Allegra. When he learned of Scotus's doctrine of the primacy of Christ he exclaimed "Voila! La théologie de l'avenir!" ("There it is! The theology of the future!"). Teilhard, like Scotus and Bonaventure, perceived Christ not only at the heart of the universe but at the heart of the material universe. He saw an intrinsic relation between Christ and the physical universe; Christ belongs to the very structure of the cosmos.[5]

Teilhard understood the science of evolution as the explanation for the physical world and viewed Christian life within the context of evolution. Evolution, he claimed, is ultimately a progression toward consciousness; the material world contains within it a dynamism toward spirit. Teilhard recognized that at all known levels of the universe there are units or "grains," which include stars, atoms, molecules, cells, people, etc. Sometimes the particles are gathered together in aggregations (e.g., a pile of sand or a galaxy of stars); sometimes they are linked together to form a crystal. While a crystal is built up by indefinitely repeating the same molecular pattern, unities in life are built up by structures ever more complex and intricately related. This "center to center" bonding of complexity, which is characteristic of all living things, captured Teilhard's attention.[6] His thought can be described in three stages of evolution. The first stage sees evolution as following an axis of increasing organization. When we consider the age of the universe (14.7 billion years), we see that there came a point when inert elements came together and formed the first living cell. All of the separate elements were there before the cell appeared, but the union of these elements caused a new entity to emerge that was more than the sum of its parts. "True unity does not fuse the elements it brings together," Teilhard wrote, rather "by mutual fertilization" it renews them; "union differentiates."[7] The movement toward complex unions means that the cell continues to reach beyond itself to find new elements and incorporate them into its unity. To describe this process Teilhard used the word "groping." The cell gropes or feels beyond itself for what it can use; indeed, the whole of life is groping since it "can advance only by endlessly feeling its way." Teilhard's word "groping" acknowledges a dependence on chance, but it also affirms that life has a direction, an orientation, a preference. Groping is "directed chance."[8] Thus, life has a finality from the beginning; it prefers increased life. Throughout its process it takes advantage of what it chances to find and puts it to its own use. Groping is a "combination of the play of chance (which is physical) and that of finality (which is psychic)."[9] In the process of groping, life tries many forms, of which some fail but others succeed. Thomas King writes, "Throughout the ages life has constructed organisms of ever greater complexity, and with this increased complexity the organism has also shown an increase in consciousness, that is, an increase of intention, of acting with a goal."[10] The second stage is one of convergent evolution headed toward a projected point of maximum human organization and consciousness, the Omega point, who is the risen Christ. In the course of evolution, the human person "emerges from a general groping of the world"; thought is born. The human person is unique, Teilhard wrote, because she or he has the ability to reflect. Thus, as human persons, we are integrally part of evolution in that we rise from the process, but in reflecting on the process we stand apart from it. The human is therefore continuous and discontinuous with evolution. Teilhard defines reflection as "the power acquired by a consciousness to turn in upon itself, to take possession of itself *as an object* ... no longer merely to know, but to know that one knows."[11] He quotes a phrase of Julian Huxley: the human person "is nothing else than evolution become conscious of itself." To this idea Teilhard adds, "the consciousness of each of us is evolution looking at itself and reflecting upon itself."[12] Thus, the human person is integral to evolution; s/he is "the point of emergence in nature, at which this deep cosmic evolution culminates and declares itself."[13] But, as King points out, "in being able to objectify the process, s/he also stands apart from it."[14] At a third level of evolution, therefore, Teilhard constructs a contemporary spirituality based on his theology of Christ, a spirituality of union with God through the world; his spirituality is based on a Christocentric vision of reality that is developed in view of convergent evolution.[15]

It is in light of the direction of evolution toward greater complexity and consciousness that Teilhard reflected on the meaning of Christ and Christian life. The process of evolution in the physical sciences may be one of cosmogenesis and biogenesis, but from the point of Christian faith it is a "Christogenesis," or a "coming-to-be" of Christ.[16] Although Teilhard was well acquainted with the biology of evolution of his time, he claimed that "there is only one real evolution, the evolution of convergence, because it alone is positive and creative."[17] He recognized that there is a unifying influence in the whole evolutionary process, a centrating factor that continues to hold the entire process together and moves it forward toward greater complexity and unity. The term "centration" refers to a convergence factor or principle. That is, as evolution proceeds toward greater complexity or divergence, it is also moving toward greater unity or convergence, that is, it is centered. Teilhard's faith in Christ led him to posit Christ, the future fullness of the whole evolutionary process, as the "centrating principle," the "pleroma" and "Omega point," where the individual and collective adventure of humanity finds its end and fulfillment, and where the consummation of the world and consummation of God converge. Christ is the personal Logos within the evolutionary process that holds things together as it moves toward unified complexity. Teilhard wrote, "We are obliged to assume the existence in the universe of a centre of universal confluence ... is not that an admirable place at which to position [or recognize] Christ?"[18] In his book *The Phenomenon of Man* he wrote, "Omega is a distinct Centre radiating at the core of a system of centres."[19] Because this center is at the core of other personal centers, it must itself possess a personality so that in the development of human personhood we begin to respond directly to the personalizing action of the center of centers.[20] If we assume Christ to be established by his incarnation at this remarkable cosmic point of all convergence, he then becomes immediately coextensive with the vastness of space.[21] In such a position, Christ is commensurate with the abyss of time into which the roots of space are driven. Through his penetrating view of the universe, Teilhard found Christ present in the entire cosmos, from the least particle of matter to the convergent human community. Christ invests himself organically with all of creation, immersing himself in things, in the heart of matter, and thus unifying the world.[22] By taking on the human form, Christ has given the world its definitive form: he has been consecrated for a cosmic function. The convergence of evolution toward greater complexity organized around a personal center meant for Teilhard that the very physical process of evolution has within it a centrating principle, which is Christ. Thus, we can say, as Augustine sought to understand the incarnation in the light of Neoplatonism and Thomas Aquinas sought to understand the incarnation in the light of Aristotle's philosophy, Teilhard sought to understand the person of Jesus Christ in the light of evolution. In his book *Christianity and Evolution* he writes,

> During the first century of the Church, Christianity made its decisive entry into human thought, boldly assimilating the

Jesus of the Gospels to the Logos of Alexandria. We cannot fail to see the logical sequel to this gesture and the prelude to a similar success in the instinct which is today impelling the faithful, two thousand years later, to adopt the same tactics— not, this time, with the ordering principle of the static Greek kosmos, but the neo-Logos of modern philosophy—the evolutionary principle of a universe in movement.[23]

Teilhard held that if "the world is convergent," he hoped to demonstrate it scientifically, and if "Christ occupies the center, then the Christogenesis of St. Paul and St. John is neither more nor less than the extension of the noogenesis (emergence of mind) in which cosmogenesis culminates."[24] Although Teilhard was influenced by the Gospel of John, especially Jesus' words "that they be one, as we are one" (John 17:21), it was St. Paul who left an indelible mark on Teilhard precisely because Paul drew an integral link between the role of Christ in the church and his role in the world. The central themes in Paul's letters, the cosmic supremacy of Christ and the plenitude of the cosmos in Christ, spoke to Teilhard of the universal significance of Christ. Christ is center of the universe who imparts to creation its harmony and cohesion and gives the world its meaning and value. As Scotus and Bonaventure maintained, creation is oriented toward Christ as its perfection and fulfillment. Christ is the purpose of this universe and, as exemplar of creation, the model of what is intended for this universe, that is, union and transformation in God. In Jesus, God accepts the cosmos definitively and irrevocably and the cosmos accepts God definitively and irrevocably. Creation is continuous and ultimately opened up to something new in Christ.

It became increasingly evident to Teilhard that if Christ is to remain himself, without diminishment, at the center of our faith, then this cosmic Christ must now offer himself for our adoration as the "evolutive" Christ who is also Christ the evolver.[25] What he sought to show was that the cosmic function of Christ was not only moral but physical as well. The Pauline phrase *omnia in ipso constant* ("in him all things consist," Col 1:17) dominated Teilhard's thought. "I find it quite impossible," he wrote in 1924, "to read St. Paul without being dazzled by the vision under his words of the universal and cosmic dominance of the incarnate word."[26] The one who is in evolution is himself the cause and center of evolution and its goal. This evolutive Christ, for Teilhard, was not distinct from Jesus but indeed was "Jesus, the center towards whom all moves."[27] It was the humanity of Jesus Christ, his life, death, and resurrection, that spoke to Teilhard of the evolutive Christ. Because Christ is both the center and goal of an evolutionary creation, Teilhard viewed Christ as a dynamic impulse within a humanity (and nonhumanity) that is moving toward greater complexity and unity, from biogenesis to noogenesis, from simple biological structures to the emergence of mind. Christ is not a static idea but a living person, the personal center of the universe. He posited a dynamic view of God and the world in the process of becoming something more than what it is because the universe is grounded in the personal center of Christ. The absolute primacy of Christ for Teilhard meant that no "spiritual substance, created or still to be created in the universe" can escape his lordship. He writes,

> It seems to me that we are living again, at a distance of fifteen hundred years, the great battles of Arianism: but with this difference, that the problem today is not to define the relations between the Christic and the Trinitary—but between Christ and a Universe that has suddenly become fantastically big, formidably organic, and more than probably poly-human (n thinking planets—perhaps millions) ... And to put it crudely (but expressively) I can see no noble or constructive outcome for the situation apart from theologians of a new Nicaea introducing the sub-distinction, in the human nature of Christ, between a *terrestrial* nature and a *cosmic* nature.[28]

Earlier, in 1945, Teilhard had written:

> We may say that the dominant concern of theologians in the first centuries of the Church was to determine the position of Christ in relation to the Trinity. In our own time the vital question has become the following: to analyze and specify exactly, in its relations, the existence of the influence that holds together Christ and the universe.[29]

Although Teilhard emphasized the cosmic role of Christ, there was still for him the one and only Jesus of history and dogmatic tradition—the historic and trans-historic Jesus—the Jesus of the

Gospel, the son of the Virgin Mary, he whose hands were pierced, the Jesus whose name comes so often in his prayers. It was Jesus, incarnate God and risen, who had made himself "evolving" that he might be "evolver." Thus Christ is not an abstract principle but a living person, as Teilhard wrote: "If it is true that it is through Christ-Omega that the universe in movement holds together, then it is from his concrete germ, the man of Nazareth, that Christ-Omega derives his whole consistence. The two terms are intrinsically one whole and they cannot vary in a truly total Christ, except simultaneously."[30] Henri de Lubac points out that, for Teilhard, the Heart of Jesus was the "fire" bursting into the cosmic milieu to "amorize" it, that is, to energize it in love. The "universal Christ" was born from "an expansion of the Heart of Jesus."[31] Because of Jesus, Christ reigns in the universe.

Teilhard's notion of a christic universe cannot be understood apart from his own spirituality centered on the humanity of Christ. The Sacred Heart for Teilhard was the Heart of Matter. It was here that God's love for the whole of the universe was revealed. Ursula King describes Teilhard's spirituality as a "heart and fire" mysticism.[32] His response to the Sacred Heart was one of love, action, and will, since this Heart was the very wellspring of his life. As he himself confessed, "the Sacred Heart has been good to me in giving me the single desire to be united with him in the totality of my life."[33] In one of his journals he wrote, "*The Sacred Heart:* Instinctively and mysteriously for me, since my infancy: the *synthesis* of Love and Matter, of Person and Energy. From this there has gradually evolved in me the perception of Omega—the universal cohesion in unity."[34] The personal love of God poured out in the crucified Christ seized the heart of Teilhard. He saw this love as a passionate love, a fire of divine love that energized and propelled this evolutionary universe toward the fullness of Christ-Omega. Robert Faricy writes,

> Through the symbol of the Sacred Heart, the Divine for Teilhard took on the properties, the form, and the qualities of a Fire capable of transforming anything and everything through the power of its love-energy. Christ, his Heart, a Fire, capable of penetrating everything—and which, little by little, spreads everywhere ... because the center of the Center of all things is the heart of Jesus, the energy that moves

the world forward into the future and that unites persons around their personal Center is love.[35]

What emerged in Teilhard, from the seed of personal devotion to the Sacred Heart, was a conviction that Christ is the center of the universe, the form of the universe, and the goal to which this evolutionary universe is directed. Christ is not merely the savior *of* the universe, as if the universe, essentially distinct from Christ, went awry and needed repair. Rather, the universe belongs to Christ. It is Christ who gives meaning and direction to the universe; Christ is the *form* of the universe. The Aristotelian philosophical terms of "form" and "matter" take on new meaning in Teilhard's thought for he looks not to the philosopher but to the Letter to the Colossians (1:16-17) where the author writes the following: "In him everything in heaven and on earth was created, things visible and invisible, whether thrones or dominations, principalities or powers; all things were created through him and for him ... and he holds all things in unity."

The belief in Christ the Center, the Heart of God, as the power of love in an evolutionary universe led Teilhard to claim that the universe is not a mere collection of parts but rather a *totum,* an organic whole. It is a *dynamic* coherence in which every element is intrinsically related to every other element. Karl Rahner said that evolutionary change occurs because of a power that comes from within the creature—the pressure of the divine acting from within. He identified this pressure from within as God who is the heart of evolutionary change, as a power that enables the creature to go beyond itself and become more than it was. The entire creation, therefore, bears a dynamic impulse toward self-transcendence.[36] The movement of self-transcendence at the heart of the cosmos reaches fulfillment in grace when the creative ground of the whole cosmic process engages in self-giving love with the free human person. The life of faith is a free response to this love, directed to us from every point in universe.

Like Rahner, Teilhard believed that the openness of the cosmos to what is new, its capacity to leap forward, leads to the emergence of intelligent beings. Within this creation process, humans are the cosmos itself come to self-consciousness. While the human person represents the highest shoot of the upper thrust of this evolution and, as microcosm, totalizes this whole, it is the human nature assumed by Christ that effects the ultimate totalization of the universe in the mysteries of creation, incarnation, and redemption. Humankind, totalizing the universe, becomes itself totalized in the incarnate Christ. Zachary Hayes writes, "We discover in a deeper sense, in what we see and hear and touch in Jesus, the divine clue as to the structure and meaning not only of humanity but of the entire universe."[37]

Teilhard's integration of the humanity of Christ and the physical universe reflects his own mystical spirit. As with other mystics, his understanding of Christ was rooted in his deep experience of God in creation. While he was aware of the doctrine of Christ formulated at Chalcedon, he did not refrain from suggesting a "third nature" of Christ based on spiritual insight and contemplation of the world. In his writings he describes this nature as follows: "Between the Word on the one side and Man-Jesus on the other, a kind of 'third Christic nature' (if I may dare to say so) emerges … that of the total and totalizing Christ."[38] Teilhard spoke of a third aspect of the theandric (divine-human) complex as "the *cosmic nature*," which, in his view, has not been sufficiently distinguished from the other two natures (divine and human). The apprehension of a third nature of Christ means that the whole physical world has a spiritual nature that attains its full consciousness and openness to God in the person of Jesus Christ. By using the term "third nature" Teilhard indicated that Christ is related organically, not simply juridically, to the whole cosmos. It is a nature contingent on the humanity of Christ but not subordinate to his divinity. Since the cosmic Christ is the resurrected Christ, this third nature, or cosmic nature, emerges from the union of divine and human natures so that it is neither one nor the other but the union of both, although it exists on the side of creation. This third nature allows us to see that Christ the Redeemer is Christ the Evolver or, in Teilhard's language, cosmogenesis is really Christogenesis.[39]

James Lyons states that Christ, in his third nature, is the organizing principle in Teilhard's evolving universe. Whereas the Alexandrian Logos was the organizing principle of the stable Greek cosmos, today, Lyons states, we must identify Christ with a "new-Logos": the evolutive principle of a universe in movement. Christ in his third nature is the prime mover of the evolving universe.[40] This third nature is reasonable if, in fact, the universe is an external embodiment of the inner Word of God, which means that there is something incarnational throughout the whole of creation. The entire cosmos, then, finds its eternal significance in the humanity of the crucified and glorified Christ, because, as Bonaventure wrote, "in his human nature he embraces something of every creature in himself when he is transfigured."[41]

It is interesting that Teilhard saw himself in the tradition of the Greek fathers, especially Irenaeus and Gregory of Nyssa. He believed his concept of Christ the Evolver should have the effect of "giving traditional Christianity a new reinforcement of up-to-dateness and vitality,"[42] presumably in the same way that the theology of the Greek fathers did. He found the Latin fathers and Western Christology on the whole too juridical; he claimed, "the Christian history of the world has assumed the appearance of a legal trial between God and his creatures."[43] It is noteworthy that writing within fifty years of the encyclical *Aeterni Patris* (1879), which mandated the theology of Thomas Aquinas for the church, Teilhard was envisioning a Christology for a whole new evolutionary world. He writes,

> The fact that Christ emerged into the field of human experience for just one moment, two thousand years ago, cannot prevent him from being the axis and the peak of a universal maturing. In such a position, Christ, wholly "supernatural" … gradually radiates his influence throughout the whole mass of nature. Since, in fact, only one single process of synthesis is going on from top to bottom of the whole universe, no element and no movement can exist at any level of the world outside the informing action of the principal center of things.[44]

Teilhard insisted that it is time to return to a form of Christology that is more organic and takes more account of physics. We need a Christ "who is no longer master of the world solely because he has been *proclaimed* to be such," he wrote, "but because he animates the whole range of things from top to bottom."[45] The greatness of Teilhard's insight into the mystery of Christ lies in the organic nature of Christ as the heart of change in the universe and Christian life as one governed by change. "If we are to remain faithful to the gospel," he said, "we have to adjust its spiritual code to the new shape of the universe. It has ceased to be the formal garden from which we are temporarily banished by a whim of the creator. It has become the great work in process of completion which we have to save by saving ourselves."[46] The Christian "is now discovering that she or he cannot be saved except through the universe and as a continuation of the universe."[47]

Teilhard was both a mystic and a prophet because he not only contemplated Christ at the heart of creation but he announced a new understanding of Christ in view of evolution, while the church adhered to a Christology of unyielding concreteness. In his own time Teilhard saw the problem with Christianity as one of increasing irrelevance. "The universe," he indicated, "is not a fixed framework upon which we have simply to project the image of Christ so that we can then quietly admire it for the rest of our days."[48] In a similar way, Zachary Hayes claims that the breakdown of a fixed cosmology by the shift from a geocentric model to a heliocentric model has led to the isolation of theology from the development of modern science. The most fundamental shift in our understanding of the cosmos, he states, is the move from the vision of a universe launched essentially in its present form by the hand of the creator at the beginning of time to a vision of the cosmos as a dynamic, unfolding chemical process, immensely large in both time and space.[49] Teilhard's basic complaint rested on an outmoded Christology formulated many centuries ago. He writes,

> Our Christology is still expressed in exactly the same terms as those which three centuries ago, could satisfy men whose outlook on the cosmos it is now physically impossible for us to accept ... What we now have to do without delay is to modify the position occupied by the central core of Christianity—and this precisely in order that it may not lose its illuminative value.[50]

His concern was to bring Christology and evolution into line with each other. As we noted before, "a Christ whose features do not adapt themselves to the requirements of a world that is evolutive in structure will tend more and more to be eliminated out of hand ... he must be presented as the saviour of the idea and reality of evolution."[51]

For Teilhard, life and thought are essentially linked. He indicated that the phenomenon of evolution in Christ is intrinsically dependent on the development of the universal Christ in our souls. His Christogenic universe invites us to broaden our understanding of Christ, not to abandon what we profess or proclaim in word and practice, but to allow these beliefs to open us up to a world of evolution of which we are vital members. He urged Christians to participate in the process of Christogenesis, to risk, get involved, aim toward union with others, because the entire creation is waiting to give birth to God's promise—the fullness of love (Rom 8:19-20). We are not only to recognize evolution but we are to make it continue in ourselves.[52] Teilhard believed that each person is a universe wherein the dust of experience is gathered into a unity. Each individual is "the incommunicable expression of a conscious point of view upon the universe. Thus the task for every person is to build her or his soul; that is, each must assemble the widely scattered elements of one's experience into a unified whole."[53] What is needed, according to Teilhard, is a personal encounter with Christ, as Christ encounters us in the vast, rich diversity of the universe. George Maloney writes:

> Knowledge of another person leading to a person-to-person relationship in love is not merely a question of conceptual knowledge but existential—not "what" but "who" one is in the being encountered. The individuality of the unique person can never be captured completely in a mental concept. The inner core of a person remains sealed off to others unless a mutual love engenders a deeper insight into who this person is. In our encounter with Christ today, it is not enough to contemplate him as the historical person revealed in the Gospels. Nor is it enough to contemplate him in the encounter of faith through grace by which we become united with him and the other members of his mystical body. We must also be able to meet Christ in the divine, continual act of creation, redemption and sanctification of the total universe … In him we discover the Absolute, the beginning and end of all unity in the cosmos. Faith gives us the eyes to see, not what God is but who God is. We must foster a steady growth of faith, a divine milieu in which Christ can be personally contacted. He is to be found at the root, the ground of our very existence.[54]

In Teilhard's view, the mystic goes about urging all reality toward the Omega, to the final synthesis that is constantly growing within him- or herself. What the mystic seeks is not an appearance of God in the world but a shining of God through his creation, a "diaphany" of God shining through the transparent world. This Christ, shining diaphanously through every creature of the universe, is encountered in a loving act of surrender in which Christ becomes the "Thou" complementing our "I." Each Christian now awakened to a new consciousness of Christ's universal presence discovers his or her own self-realization and full maturity in "being-with-Christ." Christ becomes the unifying and integrating center in creation, as each person seeks his or her self in Christ and thus in one another.

Teilhard's universe bears a striking resemblance to that of the Franciscans.[55] Like Francis of Assisi, Teilhard experienced an intimate relation with the material world and living things. Like Bonaventure, he experienced the presence of God in the universe. Bonaventure's doctrine of exemplarism, by which creation reflects God at every level, resonates with Teilhard's christic universe. For Bonaventure, the world is a mirror that reflects God. It is a stained-glass window through which the divine light shines in multicolored variety. It is a book in which we read the wisdom of God, a ladder on which we climb to God. In his *Soul's Journey into God,* Bonaventure takes his point of departure from the material world. He meditates on the material world, finding there vestiges of the Trinity. God shines through the material world in his power, wisdom, and goodness. We must contemplate God not only through creatures but also in creatures: he is in the world by his essence, power, and presence.[56] For Bonaventure, as for Teilhard, the unifying force in the cosmos is Christ.

In developing his doctrine, Bonaventure drew on the tradition of Christ as divine Logos, or the eternal image and expression of the Father. Christ is the Word who is the expression of the Father and the exemplar of all creation. All creation participates in Christ, the incarnate Word. Bonaventure contemplated Christ in all things: in material objects, in seeing and hearing, in an artisan making a table or a teacher lecturing on philosophy. He meditated on how the mechanical arts, such as weaving, agriculture, and navigation, manifest "the eternal generation and Incarnation of the Word."[57] He examined how when we think, we imitate the begetting of the eternal Word; and when we utter a word, we imitate the incarnation, or appearance of the Word in the world of sense. Like Teilhard, Bonaventure saw in matter a tendency that points to the incarnation, which is the most noble perfection of the universe. In his *Soul's Journey unto God* he articulated his dynamic thought about the soul's evolution to God: through the material world, sense knowledge, its natural faculties, as enlightened by grace, through knowledge of God as one and triune, and finally in mystical union with the crucified Christ. The journey involves a process of interiorization from the outer to the inner to the above—to union with God. In Bonaventure's view, love is the driving force and the goal of the journey.[58] Teilhard too viewed the cosmos on a journey to God in a process of divinization, which he called Christogenesis. This process takes place through interiorization and unification, from the without to the within in the Omega. Love is the force that energizes the process because love permeates the entire cosmos, that is, the affinity of being with being. Teilhard wrote, "Driven by the forces of love, the fragments of the world seek each other so that the world may come to being."[59] He identifies this energy of love with Christ, the Omega, saying, "the love of Christ is an energy into which all the chosen elements of creation are fused without losing their identity."[60] Ewert Cousins writes, "As Bonaventure saw in the individual soul, Teilhard sees in the cosmos that love is both the impetus and the term of the process of development. For Teilhard this creative love interiorizes and transforms each individual unit while uniting it to the whole."[61]

In their eschatological orientation, both Teilhard and Bonaventure were preoccupied with the future. Both mystics sought to understand the fulfillment of creation in God who is incarnate. Bonaventure's theological metaphysics allowed him to posit a relationship between Christ the center and the human person as image of God. He held that the human person as image is a creative center and open to possibilities of self-extension and development. Just as all of creation emanates from the fountain fullness of the Father and is modeled on the Son, creation finds its fulfillment in the Holy Spirit. Thus, while all of creation realizes itself as image of the Trinity by growing to its fulfillment, it is the human person whose trinitarian image is Christ, who can lead creation to its fulfillment in God. Similarly, Teilhard held that the whole of natural evolution is coming under the influence of Christ, the physical center of the

CHAPTER FIVE

RAIMON PANIKKAR AND THE UNKNOWN CHRIST

Teilhard's christic universe with its energy of convergence toward greater complexity and union is a movement toward spiritual fullness. The cosmic Christ signifies the eternal *kenosis* of God's self-sacrificing love which manifests itself within the evolutionary movement of creation and culminates by personalizing itself in the incarnation. The cosmos, or universe, is evolving toward its full realization in Christ, the Omega, who is the energizing principle underlying the process of evolution. The cosmic dimension of Christ, therefore, expresses neither his divinity nor simply the created humanity of Jesus of Nazareth; rather, the incarnation in its cosmic dimension is Christ, the God-community, into which human life and the whole cosmos are incorporated.[1] Creation will not be complete until participated being is totally united with God through Christ in the pleroma when God will be "all in all."[2]

Teilhard claimed that between cosmogenesis and Christogenesis is anthropogenesis and noogenesis. The universe is oriented toward its goal, the Omega point, through reflectively thinking human beings who emerge from the evolutionary process. The level of noogenesis is the level of mind and self-consciousness and thus is essential to the dynamism of evolution, because the human person is the growing tip of the evolutionary process and its prime influence.[3] What we think and how we act in the universe influence how the goal of the universe will be attained. Our religious nature, marked by self-transcendence, the desire for relatedness, and fulfillment, means that religion plays a vital role in shaping the human evolutionary consciousness.

It is precisely in view of religion that Teilhard thought humankind was still living in the faraway past, in the Neolithic age, as it were. He indicated that certain religious beliefs and practices of the West can be described as a form of "paleo-Christianity" which he wished to see replaced by a dynamic "neo-Christianity." Ursula King notes, however, that "no religion is free from fossilized forms." Teilhard's notion of a "fossilized Christianity" can be "applied to all past forms of religious 'other-worldliness,' to any outdated spirituality whether Eastern or Western. The need for a new road exists at present in all religious traditions."[4] Today, major changes are taking place in the world religions in terms of a new "this-worldly" orientation, she indicates. This shift accounts for the emergence of new forms of Buddhism, Hinduism, and Islam, as well as for new interpretations of Christianity.[5]

The fact that changes are taking place in the major world religions suggests that we are evolving toward a new religious consciousness, which Ewert Cousins calls a "complexified religious consciousness."[6] By this term Cousins points to a new level of consciousness in the second axial period that is different from that in the first axial period insofar as religious consciousness is no longer isolated within a culture or tradition but now coexists with other religious traditions. It is a religious consciousness, therefore, within a particular tradition in union with a consciousness of other religious traditions. The more we enter into union with other religious traditions, the greater the complexification of our own religious consciousness. The way into complexified religious consciousness is through the experience of otherness and difference, and this experience is integral to the meaning of Christ in the second axial period.

It is in view of otherness and difference—religious complexified consciousness—that the Catholic Hindu scholar Raimon Panikkar offers a profound mystical understanding of Christ in the second axial period. The son of an Indian Hindu father and a Spanish Roman Catholic mother, Panikkar wrote of himself, "I left (Europe) as a Christian; found myself a Hindu; and I return as a Buddhist, without having ceased to be a Christian."[7] Trained as a Catholic priest and steeped in the Hindu tradition, Panikkar is what Ewert Cousins calls a "mutational person," that is, "one in whom the global mutation has already occurred and in whom the new forms of consciousness have been concretized."[8] Although trained as a theologian and philosopher, Panikkar is preeminently a mystic. His insights to the meaning of Christ today flow from the depths of one who has experienced Christ—like Teilhard—in the heart of the cosmos.

Christology is essential, Panikkar indicates, because it is the name for theological reflection on the mystery of Christ. Since all Christology is a reflection on Christ, it is rooted in human experience. Christology is not a chemically pure concoction of the pure mind but has a *Sitz im Leben* that belongs also to the very interpretation of what it purports to understand.[9] Today we find a pluralism of Christologies, such as Asian, African, feminist, and liberation Christologies—a healthy sign that Christology is alive. Despite these emerging "contextual" Christologies, we are experiencing a crisis in Christianity, according to Panikkar. In some ways, Panikkar echoes Teilhard's lament, that Christianity has become "paleolithic," ancient, and out of touch with the world. The Christology of the first twenty centuries of Christian history, he writes, has been forged by the unfolding of Christian belief against the background of Hebrew religion and Greek culture. With such a ready-made Christ image, other peoples of the world have found Christ either an exotic figure or a suspicious construct.[10] Panikkar thinks that European theology, for all its brilliance and evident insight, is too much a Western historical product. What we have today, he states, "is a tribal Christology of the last two thousand years which is centered almost exclusively on its own concerns, with almost total neglect of other human religious experiences, a Christology for the internal purposes of Christians, perhaps even to conquer the world."[11] By this he means that Christology is caught up in the search for universal reason, which is, he says, an abiding Western cultural concern. In contrast, Christian self-reflection in the third millennium must be less concerned with establishing an overarching christological theory and more energized by open conversation with the multiple ways in which the mystery of Christ can be understood by various world cultures and traditions without seeking a universal explanation of the way that Christ is present elsewhere.

Panikkar holds that the encounter of traditions through multifaith (and multicultural) dialogue is crucial in the new situation of radical pluralism that confronts our world because no single religion, culture, or tradition holds a universal solution for either our theoretical or our practical human problems. The insight of Frederick Franck supports Panikkar's position:

Alone and isolated, Hinduism is threatened, Christianity is impotent, Islam is in ferment, Buddhism is dissolving, Marxism is bankrupt, secularism is self-destructing. It is not unthinkable that cross-fertilization among the traditions could reconcile the original insights of the various cultures and make the stilled voices of the sages audible once more over the abysses of time. [12]

Panikkar's own kind of radical pluralism is appealing in the manner it develops a critical stance toward all imperialistic and monistic modes of thinking and acting. No more, he states, will one religion, culture, or tradition impose itself on peoples of diverse and less powerful traditions. His vision tells us that a new wholistic experience of reality is emerging in which every tradition, religious or otherwise, can play its part in the unfolding of a new revelation where all will live in harmony and peace. This does not require the abandonment of faith, because faith is what humanity holds in common. Rather, it is by living a Christian life or, more precisely, by living Christ (especially if we profess ourselves Christian), that a new wholistic experience can develop.

When Panikkar views the Christian landscape today, he sees a rather weak religion that stakes its claim on being the "light of the world." Most Christians, he claims, are apathetic to the problems of the world and are dissatisfied with only their inner political polemics and private problems (married clergy, women priests, etc.). We have become narrow in our focus and selfish in our concerns. In a sense, we are in a new flight from the world. The problem of apathy among Christians is rampant in the First World, where consumerism dulls the human heart amidst an ecological crisis while a massive humanitarian crisis of poverty and hunger is ravaging many parts of the Third World. Panikkar points out that many people around the world live in subhuman conditions, thousands of children die every day because of human injustices, wars kill every day, and warring among religions is still very much alive. His question, therefore, is timely and critical, namely, What does contemporary Christology have to say about all this? What is the relevance of Christian belief to the burning issues of our times, and how does it all relate to Christ? "A Christology deaf to the cries of Man," he writes, "would be incapable of uttering any word of God … Christ is not a divine meteorite."[13] He is not, as Teilhard perceived, an intrusion into an otherwise evolutionary universe. Christ is not a postscript to our otherwise comfortable lives. From a Christian perspective, Panikkar states, the entire modern problematic concerning intercultural and interreligious questions hinges on the vision of Christ.[14] If we privatize and individualize Christ, how will we break through the veil of fear and distrust that shields us from other religions? Panikkar says simply, "If the mystery of Christ is not our very own ... it might as well be a museum piece."[15] For anyone who declares him- or herself a "Christian," these words must strike deeply.

Panikkar echoes Teilhard's concern that we cannot have a "pure" Christology which is a mere theological specialty severed from an underlying cosmology. Our understanding of the cosmos influences our understanding of Christ, and this knowledge of Christ cannot be a fragmented knowledge (see Phil 3:8). Hence, we must make sense of Jesus as the Christ in *this* cosmos and not the cosmos of the ancient Greeks; and this cosmos includes religious plurality. It is an interrelated cosmos with the human community emerging out of the physical processes of chemistry and biology that have led to plant and animal life. Emerging from an interrelated cosmos, the human species is interrelated as well—biologically we all share the same DNA material. Thus whatever we say of God in union with humanity, which Christ symbolizes, must pertain to the whole of humankind (and by extension, nonhuman creation).

Panikkar sees the whole body of the world's religions as expressions of the Trinity and thus a unity between Christ and creation. The silence of the Father is expressed in Buddhism; the Logos is found in Judaism, Islam, and Christianity; and the varied movements of the Spirit are present in multiple forms in Hinduism.[16] Without the Trinity there is no Christ, and without Christ there is no Trinity. If we believe Christ is redeemer, it is first of all because he is creator; Christ is first and foremost of the Trinity, and the way we conceive of Christ in the Trinity and as creator is the way we understand redemption in the mystery of Christ. If Christ has nothing to do with creation, then he also has nothing to do with redemption and thus nothing to do with humanity. Christ, united to the Father in the Spirit, is the Word and expression of the Father. As the center and Word of the Trinity, Christ is the center of creation. All things are created through Christ, the divine Word, and are restored to God through the Word incarnate. As Bonaventure wrote, "In him the first principle is joined with the last."[17]

Although Panikkar views the integral relation between the Trinity and Christ and envisions all world religions within this relationship, he still reserves a unique role for Jesus. The Christian knows Christ in and through Jesus. The title "Christ," he reminds us, is a Greek name, a translation of the Hebrew word "Messiah," which means "anointed." The generic meaning of the word, "anointed one," received a specific meaning within Judaism: the Messiah whom the people of Israel expect. The title was individualized in the person of Jesus who was recognized as the revelation of the Christ.[18] In Christian revelation and through Christian experience, the Christian discovers the Christ through whom the universe was made. The person who confesses "Jesus is the Christ" is a Christian. Panikkar says that the confession "Jesus is the Christ" is constitutively open; the predicate of the statement is a mystery, even if the subject is a concrete historical figure, the son of Mary. The Christian encounters Christ in and through Jesus,[19] however, in the community of the church and in the sacramentality of the world. This encounter cannot be one of knowledge alone, for knowledge cannot reveal one's true identity. Love, Panikkar states, is required to know a person's identity. Love is the knowledge of the mystic because love goes further than intellectual knowledge alone. We can know a person's features, his or her identification, but to know who a person is, what makes him or her unique, is to know his or her identity, a knowledge gained by the experience of love. It is love that leads us into the heart of the mystery of Christ.

Panikkar identifies the universality of the incarnation within the particularity of the Christ-event. The incarnation, he states, is not only a historical act in time and space; it is also a cultural event and intelligible only within a particular cultural setting. But the Christian incarnation is a universal human event unless we reduce Jesus Christ to a mere historical being. "Jesus is a sign of contradiction," he writes, "not because he separates me from others but because he contradicts my hypocrisy, fears, selfishness and makes me vulnerable, like he is."[20] Panikkar situates Christ within the Trinity—the Word incarnate. If we separate Christ from the Trinity, he says, the figure of Christ loses all credibility. Christ then becomes a new Socrates or some other great prophet, a great and holy figure but not divine.[21] Yet, if we sever Christ from his humanity, he becomes a platonic ideal of perfection and an instrument of dominion and exploitation of others. If we break his humanity from his historical walking on earth and his historical roots, we convert him into a mere Gnostic figure who does not share our concrete and limited human condition. In Jesus, the finite and infinite meet; the human and divine are united; the material and spiritual are one. Whoever sees Jesus Christ sees the prototype of all humanity, the *totus homo*, the full man—the new Adam.[22] Jesus Christ is the living symbol of divinity, humanity, and the cosmos (material universe) united as one—a "cosmotheandric" symbol.[23] Panikkar claims that we have lost sight of the mysticism of Jesus Christ, which modern Christology has ignored. He asks, "Is Christian faith founded on an historical book or a personal experience?" If we answer, "historical book," then we are talking about a museum piece. If we answer "personal experience," then we have to begin to make sense of Christ in the face of religious pluralism.

Christophany

Christ is the central symbol of all reality, according to Panikkar. In Christ are enclosed not only "all treasures of the divinity but also hidden 'all the mysteries of Man' and all the density of the universe. He is not the symbol *for* reality but the symbol *of* reality."[24] As the center of the Trinity, he is the symbol of the entire divinization of the universe. Like Teilhard, Panikkar relies on the following Pauline theme: "For in him the whole fullness of divinity dwells bodily" (Col 2:9). Thus Panikkar concludes, "Jesus is Christ but Christ cannot be identified completely with Jesus."[25] "Christ infinitely surpasses Jesus."[26] Christ is not only the name of a person, he claims, but the reality of our own lives; that is, Christ does not belong only to the Jesus of history but the living human person united with God at the heart of the universe. Thus, he points to a deep inner center in the human person with the capacity to manifest Christ—what he calls "christophany."[27] Christophany makes sense only within a trinitarian insight; that is, only in light of Christ the divine Word who is the Logos of the Father and thus the creative principle of God does every aspect of creation express Christ. This idea is consonant with Bonaventure's doctrine of expressionism by which everything in creation expresses the Word of God.

The term christophany has as its root the *phaneros* of the Christian scriptures: a visible, clear, public manifestation of a truth. "Christophany stands for the disclosure of Christ to human consciousness and the critical reflection upon it";[28] each person bears the mystery of Christ within. The first task of every creature, therefore, is to complete and perfect his or her icon of reality,[29] because "Christ is not only the name of a historical personage but a reality in our own life" (Phil 2:7-11).[30] Christ is the symbol of our human identity and vocation which, in its acceptance and fulfillment, is the union of all created reality in the love of God. The Greek word "symbol" comes from the verb *symballein*, which means "to throw together." Louis Chauvet describes the meaning of symbol as follows:

> If one construed it transitively, one would translate it, according to the context, as "gather together," "hold in common," or "exchange." The substantive *symbole* designates the joint at the elbow or knee and, more generally, the whole idea of conjunction, reunion, contract, or pact. The ancient *symbolon* is precisely an object cut in two, one part of which is retained by each partner in a contract. Each half evidently has no value in itself and thus could imaginatively signify anything; its symbolic power is due only to its connection with the other half. It (can thus be described) as an agreement between the two partners which establishes the symbol; it is the expression of a social pact based on mutual recognition and, hence, is a mediator of identity.[31]

The symbol introduces us into a realm to which it itself belongs, that is, an order of meaning in its radical otherness.[32] A symbol retains its value only through the place that it occupies within the whole; an element becomes a symbol only to the extent that it represents the whole, from which it is inseparable.[33] Panikkar uses the word "symbol" to "express an experience of reality in which subject and object, the interpretation and the interpreted, the phenomenon and its noumenon, are inextricably linked."[34] Gustave Martelet writes that "Christ is the fullness of sym-bol, for in his incarnation, death and resurrection, he is the *fully actualized gather-together of God and human*, of the eternal and passage of time."[35] Because Christ is the symbol of human life and the reality of what we are to become, we must discover the Christ within us as the very meaning of our lives. Christ is not a definition or formula, a conundrum of multiple natures to be solved by human logic. Christ is a personal, living reality. We may pray to Christ in a formula of prayer, but do we live Christ? Panikkar states that the task of Christians (today)—"perhaps even our kairos—may be the conversion—yes, conversion—of a tribal Christology into a christophany less bound to a single cultural event."[36] As he writes, "The whole of reality could be called, in Christian language, Father, Christ, Holy Spirit—the Font of all reality, reality in its act of being (that is, its becoming the existing reality which is 'the whole Christ') not yet fully realized, and the Spirit, the wind, divine energy that maintains the perichoresis in movement."[37] By seeing the whole reality in the Christ mystery, he claims, every being is a christophany, "a manifestation of the christic adventure of the whole of reality on its way to the infinite mystery."[38]

Panikkar assures us that christophany "does not claim to be universal; it is not an attempt at a new Christian imperialism, and it does not offer a universal paradigm."[39] His concern rather is directed toward the "post-Christian," those contemporary Christians who think that in order to be ecumenical, open, tolerant, and fully Christian, they have to dilute their "christianness" and water down their fidelity to Christ.[40] The idea of christophany, he claims, does not totally depart from Christology but underscores (1) a more passive attitude of receiving the impact of Christ over and against a more aggressive search by human reason for intelligibility; (2) a reintegration of the Christ figure into a cosmological vision that is relevant to our age; (3) a thematic integration of the homeomorphic equivalents of what Christians call Christ, that is, aspects of Christ that may be found in other religions under the language and categories of that tradition. He assures us that christophany does not abolish the Christological tradition of the past two millennia; rather, it is continuous in that it builds on the insights of traditional Christology, but it is also discontinuous with the tradition insofar as the significance of Christ is broadened beyond the community of the baptized. The notion of Christ includes both the figure of the historical past and the reality of the present. Christophany, Panikkar states, is not an exercise in Christian solipsism. It is a type of Christian reflection that takes into account the self-understanding of the other traditions of the world;

it is the fruit of a dialogue with other religions as well as an interpretation of the Christian tradition itself. One could say that
Panikkar's christophany finds support in Teilhard's maxim, union
differentiates. The more one is union with another, the more one
becomes one's self, because it is the self that is the very basis of
union.[41]

Panikkar affirms that Christ is our destiny and the way to that
destiny. Without Christ, creation finds neither purpose nor fulfillment, and without creation, Christ does not exist. Incarnation and
creation are inextricably linked in the communion of trinitarian
love. Panikkar states, "God is mystery and we too exist within this
mystery."[42] We are caught up in the "I-Thou" relationship of the
Father and the Son in such a way that, like Jesus, we too are the
"you" of the Father.[43] Although we are not the Son of God, we are
"little words" of God, finite expressions of the infinite love of the
Father expressed in the Son. We, too, therefore are entwined in the
infinite "Thou-ing" of the Father. Every human person manifests
Christ because like Christ we are the children of God and thus are
part of the divine outpouring of the Father. The Father is the
source of my being, the mysterious Fountain of Being, so that
everything that I am, including what I define as "mine," is pure
gift. Everything is grace. "Although I am not the ground [of my
existence]," Panikkar writes, "neither does that ground exist outside of me ... In other words, the *Grund* is not an 'other,' a non-
I but a 'thou,' an immanent transcendence in me—which I discover
as the *I* (and therefore as my I)."[44] Cousins writes that "Panikkar
opens up for us our own relationship to the Father as being Christ-
centered, that we as the children of God are part of the divine outpouring of the Father ... This inner fountain or divine indwelling
in the human person is an invitation to manifest the divine. It is
becoming another Christ."[45]

Panikkar summarizes his doctrine in what he calls a "christo-
phanic nonary," nine points or *sutras* that outline the essential
tenets of christophany.[46] An abbreviated version of these points is
listed here:

1. *Christ is the Christian symbol of all reality.* Christ is the symbol of all reality, the symbol of the entire divinization of the
   universe.

2. *The Christian knows Christ in and through Jesus.* In Christian revelation and through Christian experience, the Christian discovers the Christ through whom the universe was made. This discovery is one of personal experience in the community of believers, but it does not exhaust the mystery of Christ.

3. *Christ's identity is not his identification.* A person may be objectively identified through a series of facts, but love is required to know a person's identity. To know Christ is to experience Christ and to reflect on that experience.

4. *Christians do not have a monopoly on the knowledge of Christ.* Reflection on Jesus and the experience of Christ by Christians does not exhaust the riches of the reality of Christ. The "unknown Christ of Hinduism," for example, is *a fortiori* unknown to Christians, and Hindus do not need to call it by that name. They call "it" by different names, although Christians are to recognize in those insights lights of the same mystery, so that there are other aspects of Christ unknown to them. Christ surpasses all understanding.

5. *Christophany is the overcoming of a tribal Christology.* Christians must convert a tribal Christology into a less sectarian christophany. A contemporary Christology has to incorporate a study of how the mystery of Christ manifests itself in other religions.

6. *The protological, historical, and eschatological Christ is one and the same reality distended in time, extended in space, and intended for us.* Christophany makes sense only within a trinitarian insight. Christ is the full manifestation of the Trinity, which is a perichoresis [mutual indwelling] of distinct persons [and thus personalities] in love.

7. *Incarnation is also inculturation.* The incarnation, as Christians understand it, is not only a historical act in time and space; it is also a cultural event and only intelligible within a particular cultural setting.

8. *The church understands itself as the locus where incarnation takes place.* The primordial meaning of the word "catholic" is the church that exists throughout the universe. God created the world in order to divinize his creation by letting it become his own body, whose head is Jesus Christ, and we are the body extended.

9. *Christophany is the manifestation of the mysterious union of the divine, human, and cosmic "dimensions" of reality.* Jesus Christ is the prototype of all humanity. In Christ the finite and infinite meet; the human and the divine are united. In him, the material and the spiritual are one, to say nothing of masculine and feminine, high and low, heaven and earth, time and eternity. Christ is a "cosmotheandric" reality, the reconciliation of the divine and the universe, who calls us to reconciliation, that is, overcoming otherness.

Reflection on Christ in a global context should not only affirm Christ for Christians but also compel Christians to live Christ more fully. Only the Christian who is living Christ can see Christ beyond the boundaries of institutionalized Christianity and search the depths of Christ in the brother or sister of another religion or culture. For the one who lives in Christ comes to see that "Christ is that central symbol which embodies the entire reality."[47]

The Role of the Spirit

Panikkar's insight into the universal Christ who is the meaning and goal of an evolutionary universe compels us to ask, How does the fullness of Christ today come to be realized in creation? The answer is simply, through the power of the Spirit. It is the Spirit sent by Jesus Christ that enkindles the mystery of Christ within us and among us. Panikkar looks to the Gospel of John to realize that the power of Christ is in the power of the Spirit. It is significant, he says, that Christ does not impart the Spirit during his earthly life but the Spirit is sent upon his death and resurrection. "Unless I go away," Jesus said, "the Paraclete will not come to you; but if I go, I will send him to you" (John 16:7). In the Gospel of John we realize that Jesus must go away in order for the Spirit to come. The departure of Jesus is the arrival of the Spirit. The going away of the earthly Jesus is the beginning of the resurrected Christ made possible by the Spirit. One must wonder why Jesus could not stay to finish the work he had begun. Instead, he promises the Spirit to his followers at a moment when his whole mission borders on the brink of disaster. Panikkar writes,

The future does not seem bright, his followers will be perse-
cuted. The Master is about to leave without having finished
hardly anything while almost abandoning his disciples. The
people have abandoned him because it has become too risky
to follow him; the synagogue declares him a heretic, indeed
blasphemous; the political representatives despise him; and
his "own" do not understand him. He has not left them anything durable, no institution; he has neither baptized nor
ordained, much less had he founded anything. He has left
both the Spirit and himself as a silent presence in the
Eucharistic act. He has sent his disciples as lambs among
wolves and refuses to change tactics even at the end: wolves
are still roaming about. He promises his followers only one
thing: the Spirit.[48]

There is something worth noting about Jesus' imminent departure at a time when his mission was just getting under way and his
disciples were increasing in number. Jesus does not stay around to
complete his earthly work; rather, he promises the Spirit to his disciples, for it is they, he indicates, who will complete his work.
"Whoever believes in me," he said, "will perform even greater
works, because I am going to the Father" (John 14:12). Imagine
the utter amazement and confusion of the disciples during the final
days of Jesus' earthly life. Indeed, they feared the loss of their master and teacher. How could they perform greater works than Jesus?
The answer is simply, the power of the Spirit. Through the power
of the Spirit, Jesus indicated that his disciples could do greater
works than he, for the Spirit does new things; and only in the Spirit
can this world move forward toward the fullness of life in God.
Thus, Jesus said to his disciples, "It is good that I am leaving you."
Otherwise, Panikkar writes, "we would make him king—that is, an
idol—or we would rigidify him into concepts, into intellectual con-
tainers. We would turn his teaching into a system, imprison him
within our own categories and suffocate the Spirit."[49] Rather, Jesus
knew that it was good that he leave, that he had not come to
remain but to remain in us in the most perfect form, not as a more
or less welcome guest foreign to us but in our very being. This is
the meaning of the Eucharist. This is the work of the Spirit and the
meaning of Christ: "I am with you always, until the end of time"

(Matt 28:20). Jesus leaves so that the dynamism of life will not be reduced to an arid dualism. His departure does not signify the departure of God from the world but the release of God into the world. This release of God is the power of the Spirit who permeates our lives in the name of Christ and reconstitutes the body of Christ in a new way, the way of unifying love amidst the delightful diversity of God's creation.

Jesus departs from this world to become something more for the world—the Christ—the beloved Word and Wisdom of the Father. The Spirit is what brings Christ to the fullness of life in creation: "Unless I go," Jesus said, "the Advocate will not come to you; but if I do go, I will send him to you. And when he comes ... He will glorify me" (John 7:14). Jesus departs from the world to become the fullness of life for the world, the fullness of what he has been from eternity, the beloved of the Father. What God had intended for all eternity—the Christ—now becomes concrete reality in the person of Jesus, who in his death and resurrection is anointed as the Christ—the incarnate Word risen and glorified. Because the humanity of Jesus is our humanity, what happens in Jesus is our destiny as well—transformation and union in God.

Jesus indicated by his radical departure from earthly life and the sending of the Spirit that the whole creation has yet to become the fullness of Christ. He said to his disciples, "As the Father has loved me, so I have loved you: Remain in my love" (John 9:12). Panikkar indicates that we are asked to *continue* the incarnation, to allow the Word of God to take root within us, to allow it to become enfleshed in us. The incarnation is not finished; it is not yet fully complete, for it is to be complete in us. As he states, the continuation of the incarnation "liberates us from living in a merely historical and temporal universe and makes us conscious of our divine dignity."[50] We are called to continue that incarnation toward the new incarnation, the fullness of Christ, which is all humanity and creation bound in a union of love. Salvation consists in reaching our own fullness, in sharing the divine nature, because nothing finite or short of God can satisfy the human longing for God. The human person and the entire creation can reach that fullness because at the very root of creation is the mediator, the Christ, who being begotten by the source and origin of all divinity not only creates everything but divinizes everything by the grace of the divine Spirit. The adventure of the entire creation is the divine life

reconciliation; it is the very process of peacemaking and reconciliation, a process that occurs in the acts of living together, working together, and praying together. It is the overcoming of otherness and the journeying toward union. Such dialogues, he indicates, are powerful means to correct biases, erase deep-seated hatreds, heal ancient wounds, and can help forge a new way of life.[55]

Panikkar himself distinguishes between interreligious dialogue and intrareligious dialogue. The first confronts already-established religions and deals with questions of doctrine and discipline. It is a dialogue of dialectics, which pursues truth by trusting the order of things, the value of reason, and weighty arguments. Dialectics depends on the optimism of reason and maintains that truth can be found by relying on the objective consistency of ideas. Dialogue, Panikkar states, does not seek primarily to be duo-logue, a duet or two sets of ideas, which would still be dialectical, but a *dia-logos*, a piercing of the *logos* to attain a truth that transcends it.[56] In this respect, true dialogue must be *intra*religious dialogue. This dialogue is the optimism of the heart; it believes it can advance along the way to truth by relying on the subjective consistency of the dialogical partners. It does not begin with doctrine, theology, and diplomacy. It is "intra," which means, as Panikkar explains, if I do not discover in myself the terrain where the Hindu, the Muslim, the Jew, and the atheist may have a place— in my heart, in my intelligence, in my life—I will never be able to enter into a genuine dialogue with him. As long as I do not open my heart and do not see that the other is not an other but a part of myself who enlarges and completes me, I will not arrive at dialogue.[57] True dialogue is embracing the other in the midst of unanswered questions. An embrace, Miroslav Volf writes, begins with opening the arms. "Open arms are a gesture of the body reaching for the other. They are a sign of discontent with my own self-enclosed identity and a code of desire for the other. I do not want to be myself only; I want the other to be part of who I am and I want to be part of the other."[58] A self that is "full of itself" can neither receive the other nor make genuine movement toward the other.[59] Open arms signify that I have "created space in myself for the other to come in and that I have made a movement out of myself so as to enter the space created by the other."[60] Volf indicates, however, that one does not stop at the embrace, for the purpose of the embrace is not to make two bodies one; it is not meant to dissolve one body into the other. If the embrace is not to cancel itself, therefore, the arms must open again; this preserves the genuine identity of each subject of the embrace.[61] Nor should we try to understand the other if we are to preserve the "otherness," the genuine identity of the other in the embrace. If we try to understand the other on our own terms, we make the other into a projection of ourselves or try to absorb the other into ourselves. A genuine embrace entails the ability not to understand but to accept the other as a question in the midst of the embrace, and to let go, allowing the question of the other to remain mystery.[62] Volf's phenomenology of embrace helps us realize that real intrareligious dialogue begins in myself, recognizing my own inner poverty and need for otherness; thus it begins in prayer. It is more an exchange of religious experiences than of doctrines, or, rather, it is precisely through an exchange of religious experiences that we can fruitfully share doctrine as well. If one does not start from a foundation of inner poverty and prayer, no religious dialogue is possible; it is just idle chatter.[63]

Dialogical dialogue requires certain attitudes and dispositions, which include a deep human honesty, intellectual openness, and a willingness to forgo prejudice in the search for truth while maintaining "profound loyalty towards one's own tradition."[64] It begins with the assumption that the other is also an original source of human understanding and that, at some level, persons who enter the dialogue have a capacity to communicate their unique experiences and understandings to one another.[65] Such dialogue can proceed only on the basis of trust in the "other as other," that is, without making the other into an image of oneself. It calls for a "cosmic confidence" in the unfolding of reality itself.[66] Such confidence, however, does not arise from intellectual understanding alone but from the maturity of a deep inner spirit rooted in God. Without spirituality as the foundation for dialogue, fear and distrust easily disrupt the desire for dialogue so that confidence in the other becomes suspicion and defense of one's position over and against the other. Dialogue is subverted and reduced to conquering and converting the other to one's position. Panikkar affirms that dialogue should not—indeed cannot—assume a single vantage point or higher view outside the traditions themselves. The ground for understanding needs to be created in the space between the traditions through the praxis of dialogue.

The starting point for dialogical dialogue, according to Panikkar, is the intrapersonal dialogue by which one consciously and critically appropriates one's own tradition. Without this deep understanding of and commitment to one's own tradition, there are simply no grounds for the dialogical dialogue to proceed. One also needs a deep commitment and desire to understand another tradition, which means being open to a new experience of truth since "one cannot really understand the views of another if one does not share them."[67] This is not to assume an uncritical approach to the other tradition so much as a willingness to set aside premature judgments that arise from prejudice and ignorance, the twin enemies of truth and understanding. Dialogical dialogue in Panikkar's view is primarily the meeting of persons; the aim is "convergence of hearts, not just coalescence of minds."[68] In the encounter, each participant attempts to think in and with the symbols of both traditions so that there is a symbolic transformation of experiences. Both partners are encouraged to "cross over" to the other tradition and then "cross back again" to their own. In so doing, they mutually integrate their testimonies "within a larger horizon, a new myth."[69] Not only does each begin to understand the other according to the other's self-understanding, but there is growth and dynamism in the manner that each tradition understands itself.[70] The impact such dialogue would have on mission cannot be overemphasized and would require a lengthy discussion beyond the scope of this text. Mission as dialogue, however, expresses the new understanding of Christ in the twenty-first century.

Dialogical dialogue challenges the notion that religions are closed and unchanging systems. It assumes instead that one is able to enter into and experience the symbolic world of the other and, on the basis of such experience, integrate it into one's own tradition. Panikkar's notion of "homeomorphic equivalence" is designed to respond to this challenge. The term "homeomorphic equivalence" refers to a "correlation of functions" between specific beliefs in distinct religious traditions. This is the method that Panikkar follows in his work *The Unknown Christ of Hinduism*. Christian belief in Christ and the Vedanta Hindu understanding of Isvara are notably distinct; however, certain correlations emerge once both Christ and Isvara are interpreted according to their respective functions within their own traditions. Christ's role as the one and only mediator between God and the world is not without meaning for the Vedanta Hindu, who would call this Isvara but understand it differently because of different conceptions of a personal creator God (Yahweh) and the impersonal noncreator Brahman.[71] The discovery of functional similarities between religions can arise only from the praxis of dialogical dialogue. The crossing over into the religious world of the other can be a moment of revelation or enlightenment in which the encounter between different religious or cultural worlds reaches a new stage of being. Such dialogue, Panikkar indicates, is not only a growth in human consciousness but "the whole universe expands."[72]

The expansion of human consciousness through dialogic dialogue reveals the mystery of Christ as a coat of many colors. The fullness of the body of Christ is the diversity of its members. The greater the diversity of the body of Christ, the greater the fullness of Christ. This new understanding of Christ in the second axial period is the emergence of Christ as the meaning of the universe within a new type of complexified religious consciousness. The insights of Panikkar have shed new light on the *mystery* of Christ. He reminds us that in an age of humanitarian and ecological crises, there is no room for a triumphalist religion, as if any one religion might conquer the world. In such a religion, we might say, Jesus is crucified over and over again—nailed to the cross, stripped of his garments, and left to die alone. To be a Christian in this new age is to see in Jesus the symbol of all humanity and thus to help shine this light of Christ in the world. Christ must be raised from the dead if this world is to find its true unity in God—a resurrection not of the man Jesus, which has already taken place—but a resurrection of our own lives. Through the cross we must let go and die into love so that the power of the Spirit may raise us to new life, the life of Christ. This is the foundation of a christic universe, as Panikkar describes it, and until we can recognize ourselves as integral to the mystery of Christ, Christ is confined to the local tribes of Western culture.

CHAPTER SIX

THE TRANSCULTURAL CHRIST: THOMAS MERTON AND BEDE GRIFFITHS

As we consider the mystery of Christ within the complexified religious consciousness of the second axial period, we see that understanding Christ in an evolutionary universe is less an exercise of the intellect than of the heart. It is a way of coming to understand a fuller, richer expression of humanity within the unfolding diversity of creation. Knowledge therefore must always become a step toward ever deeper, richer love and transforming union with the God of love.[1] While Christology of the first axial period tends to be rational and demonstrative, the complexification of Christ in the second axial period through the evolution of human consciousness is more mystical, affective, and relational (without denying the rational). It is based on participation of the knowing subject in spiritual realities, which is not simply engaging in intellectual thought but also being open to the gifts of the Holy Spirit and advancing in the spiritual life; that is, praxis rather than theory. The emergence of a universe richly complex and diverse means that Christology today must be at home in complexity and diversity and thus in the realm of mystery. In light of Teilhard de Chardin and Raimon Panikkar, I suggest that theology in the second axial period must be born of mystical insight. The theologian is the mystic who leads us beyond the boundaries of intellectual reason into the mystery of God hidden in the heart of the world.

If mysticism is indeed at the heart of theology in the second axial period then spirituality is also integral to the theological task. Dialogic dialogue may be a new way to live the mystery of Christ today, but it is spirituality that engenders a fruitful dialogue. As I indicated in the last chapter, spirituality is the root of dialogical dialogue. Without a deep, personal relationship to God (or the divine), dialogue may easily collapse into opposition marked by distrust and defense of doctrine. In this respect it is not surprising to find that our two final guides to the meaning of Christ in the second axial period are men of profound spirituality. Thomas Merton and Bede Griffiths were monks of the Benedictine tradition, as well as mystics and dialogue partners with Eastern religions. The path to dialogue for each developed from a deep, intimate relationship with Christ, not unlike many mystics of the Middle Ages. Merton's transcultural Christ is a more contemporary approach to union with Christ. The deeper one enters into the mystery of Christ in one's life, the more one finds Christ at the heart of the universe; everything in some way speaks of Christ, who transcends all boundaries. The one who lives in Christ, therefore, lives in unity (see Gal 3:28). Similarly, Bede Griffiths speaks of dialogue in the cave of the heart, that deep inner center in every human person where the divine Word dwells. To be in union with Christ is to unite with others in whom Christ lives. It is to act from a deep inner center of love that transcends divisions and thus unites with others through an embrace of crucified love.

While Merton and Griffiths differ in their approaches to living the mystery of Christ, they both represent the monastic approach to Christ in which union with the humanity of Christ leads to the divinity of Christ. That is, there is no path to God apart from the human person. In this respect, dialogic dialogue is not simply an effort to foster unity among world religions; it is a way to union with God in and through the mystery of Christ, which includes our brothers and sisters of different religions. Although Merton was a more prodigious writer than Griffiths, both thinkers have left behind a legacy that illuminates the meaning of a complexified religious consciousness centered in Christ.

Thomas Merton

Thomas Merton (1915-1968) was born in Prades, France, to artists Ruth and Owen Merton. His early years were spent in the south of France; later, he went to England, first to private school and then to Cambridge. Both of his parents were deceased by the time Merton was a young teen, and he eventually moved to his grand-

parents' home in the United States to finish his education at
Columbia University in New York City.[2] While a student there, he
completed a thesis on William Blake, who was to remain a lifelong
influence on Merton's thought and writings. But Merton's active
social and political conscience was also informed by his conversion
to Christianity and Catholicism in his early twenties. He worked
for a time at Friendship House under the mentorship of Catherine
Doherty and then began to sense a vocation to the priesthood. In
December 1941, he resigned his teaching post at St. Bonaventure
College, in Allegheny, New York, and journeyed to the Abbey of
Gethsemani, near Louisville, Kentucky. There, Merton undertook
the life of a scholar and man of letters, in addition to his formation
as a Cistercian monk. As a Trappist monk, Merton wrote numerous books on Christian spirituality, monasticism, and social commentary; and he was a forerunner in popular interreligious dialogue
in the twentieth century.

Merton, like Teilhard and Panikkar, believed that contemporary
Christianity was in a diaspora, no longer the dominant religious
institution of Western culture but now scattered and enervated
throughout the world. He also believed that Christianity was
undergoing a process of detachment from its inherited Western
forms (with its Greek philosophical trappings) and was being summoned to a form of transculturalization.[3] It is interesting to note
that he did not view this new emerging form of Christianity as a
"post-Christian" religion but precisely one centered in Christ in
whom he saw the power of the risen one at work in the world. As
William Thompson points out, Merton's Christ-mysticism is integrally related to his transcultural consciousness. Merton believed
that only in union with Christ, who is the fully integrated person,
can one become transpersonal, transcultural, and transsocial. Only
in union with Christ, the One, can a person be united to the many
since, as Word and center of the Trinity, Christ is both the One and
the Many. One who lives in Christ, therefore, lives not in division
or isolation but in the unity of love.

Merton's early secular life, a constant movement in the bohemian
lifestyle of artists and writers, led him to realize that individuals can
exist in a superficial, exterior state that is contrary to the love of
God and neighbor. The journey toward God requires an inner
renunciation of all relationships to persons and objects that are
based on a superficial understanding of reality. The journey inward

involves pain and the loneliness of letting go but also opening one-self to the vulnerability of growth and change. This pursuit eventually leads to a deeper understanding of who we are in our truest inner person, and at the same time brings us closer to God and to others. The direction inward is the only genuine way to find the healing unity of love, the freedom found in real autonomy, and the true happiness experienced in understanding the "true self" and God.

Merton held that the spiritual life is the foundation of a new consciousness of Christ. The monk, who is the solitary one, freely abandons oneself to God for the sake of the fullness of life. The solitary oneness with God is that encounter of self within where God reveals Godself and loves Godself in the uniqueness of one's created being. In his *Conjectures of a Guilty Bystander* Merton writes,

> From that secret and unspoken unity in myself can eventually come a visible and manifest unity of all Christians. If we want to bring together what is divided, we cannot do so by imposing one division upon the other or absorbing one division into the other. But if we do this, the union is not Christian. It is political and doomed to further conflict. We must contain all divided worlds in ourselves and transcend them in Christ.[4]

Merton provided a path to oneness with God through the differentiation of the "false self" and the "true self." In his *New Seeds of Contemplation* he wrote that the seed of our identity lies in God; thus it is only in God that we can find who we are and what we are created to be. The further I am from my true self, he said, the self that God created me to be, the more wrapped up I am in my false self, the self I think I need to be and the one that is farthest from God.[5] This self, because it is in darkness, demands all my time and attention, turning me away from my true self embedded in the web of creation. The emptiness of the false self leads us to a spiritually impoverished state distant from our inner life, which is rooted in unity and love. This realization awakens us to the importance of shedding our illusory existence in order to embrace a greater fulfillment in God. This is the essence of Christian self-denial, according to Merton: "It is [in] the denial of our unfulfillment, the renunciation of our own poverty, that we may be able to plunge freely into the plenitude and the riches of God ... without looking back on our own nothingness."[6] In order to be able to look beyond our emptiness, we have to let go of exterior attachments and trust in God's mercy.

Merton's journey into the transcultural Christ is an inward journey to the true self. He suggests that the fulfillment of our deepest humanity lies within us, not outside ourselves. The journey inward takes place beyond the ego, in a personal, mystical struggle within an inner cloud of darkness. He calls us to shed our false selves by breaking down the barriers of isolation that have formed around our lives. He speaks of a new birth within, which involves struggle, yet a birth to a more liberating, creative power of life. He states, "There is in us an instinct ... for renewal, for a liberation of creative power ... [It] tells us that this change is a recovery of that which is deepest, most original, most personal in ourselves. To be born again is not to become somebody different, but to become ourselves."[7] This is the basis of Merton's transcultural consciousness, a "final and complete maturing of the human psyche on a transcultural level."[8] "What Merton has in mind," Thompson states, "is the emergence of a person of such inner calm and personal and cultural detachment that she is capable of recognizing and perspectivizing the genuine values present in every person and every culture."[9] The transcultural person does not repudiate society but lives and acts within it from a new level of consciousness, a new reality that embraces all people and all life. This new consciousness opposes the isolated self forged by technological society, which forces us to master reality and isolate ourselves from the masses by becoming wholly self-sufficient. In recognizing the reality of the human situation, we embrace a true humility, which Merton says must underlie all relationships. Humility discards all that is false and sees reality as it is, the opposite of the distant manipulation of the world by the ego. We learn that our lives have real meaning only insofar as we have oriented our hearts toward God as the source and end of reality.

Merton's search for the true self in God is similar to Bonaventure's "soul's journey." The journey to the true self *is* the journey to God, and the journey to God *is* the journey to the true self, in which we find ourselves as relational beings, the image of God.[10] As Thompson writes, "the Mertonian self is co-constituted by a transcendent Self" so that "the self finds its identity not in the individual self as a separate, limited and temporal ego, but in Christ, or the Holy Spirit within this self."[11] Thus, for the self to be true to what it is, a relational being, it must be in union with another; as Teilhard reminded us, union differentiates, and the more we are in union with others, the more we are ourselves.

For Merton, solitude is the means of coming to a fuller awareness of oneself. The state of being aware of God and conscious of our true selves is not a single fantastic event but a process of growth and continuous inner renewal. The solitary life is not necessarily physical separation from the world but a life of constant self-denial and rebirth in Christ. In solitude, we establish our foundation in God and preserve it through contemplative prayer. This experience allows us to remain open to God, to our inner selves, and to all of creation. Merton explains, "True solitude is the home of the person, false solitude the refuge of the individualist. The person is constituted by a uniquely subsisting capacity to love—by a radical ability to care for all beings made by God and loved by Him. Such capacity is destroyed by the loss of perspective."[12] In the inner life, the human condition is healed by the transformative action of God. In solitude, the person is at once alone yet in communion with all of humanity in the body of Christ. Merton states, "He is one in the unity, which is love. He is undivided in himself, because he is open to all. He is open to all because that is the source of all, the form of all and the end of all, is one in him and in all. He is truly alone who is wide open to heaven and closed to no one."[13] The growth of transcultural consciousness comes from discovery of self in God:

> The deep "I" of the spirit, of solitude and of love, cannot be "had," possessed, developed, perfected. It can only *be*, and *act* according to deep inner laws which are not of man's contriving but which come from God ... This inner "I" who is always alone is always universal: for in this inmost "I" my own solitude meets the solitude of every other man and the solitude of God. Hence it is beyond division, beyond limitation, beyond selfish affirmation. It is only this inmost and solitary "I" that truly loves with the love and the Spirit of Christ. This "I" is Christ himself, living in us: and we, in Him, living in the Father.[14]

The solitary, therefore, is not called to leave society but to transcend it; not to withdraw from the fellowship with other humans but to renounce the appearance, the myth of union in diversion in order to attain to union on a higher and more spiritual level.[15] Humanity can become fully alive only in such a state of solitude. Humans are truly alive when they experience a fullness of intelligence, freedom, and spirituality within themselves, when they can recognize and interpret the will of God. We are authentically free when the brokenness and divisions that marked the false self are healed. This is the only way for us to "re-unite" with our true inner self, with God, and with others. The importance of the true inner self led Merton to reflect on the dangerous impact of the technological culture, which, in his view, could lead to a "centrifugal flight that flings the human person, in all his or her compact and uncomprehending isolation, into the darkness of outer space without purpose and without objective."[16] Without solitude, he claimed, there can be no compassion because the human person becomes lost in the wheels of a social machine and is no longer aware of human needs as a matter of personal responsibility.[17] To explain the problem of isolation, Merton drew on the image of the body of Christ, to which all humanity is called to be united. Society, he states, is a body of "broken bones"; it is "dis-tracted," broken by the unnatural forces of selfishness and greed. While the body of Christ is to be united by love, modern society fragments the body by luring people into isolation and loneliness. The inner self becomes neglected by the corrupted will of the ego-self. Only through the crucified Christ, he indicates, can the bones be reset.[18]

Merton suggests that a transcultural psyche, which is the fruit of a mature spiritual life, is the basis of a fully integrated person who attains a deeper, fuller identity as person and hence can "rise above," so to speak, cultural fragmentation and religious difference. In his *Contemplation in a World of Action* Merton writes,

> The man who has attained final integration is no longer limited by the culture in which he has grown up … He accepts not only his own community, his own society, his own friends, his own culture, but all mankind. He does not remain bound to one limited set of values in such a way that he opposes them aggressively or defensively to others. He is fully "Catholic" in the best sense of the world. He has a unified

vision and experience of the one truth shining out in all its various manifestations, some clearer than others, some more definite and more certain than others. He does not set these partial views up in opposition to each other, but unifies them in a dialectic or an insight and spontaneity into the lives of others. The finally integrated man is a peacemaker, and that is why there is such a desperate need for our leaders to become such men of insight.[19]

Thompson points out "that Merton's view of the transcultural personality provides us with the creative underpinnings of a deeper view of Christology itself ... and with a new Christ-vision capable of fostering in the Christian West our transcultural consciousness."[20] Through the continuous practice of prayer, self-reflection, meditation on the Word of God, and contemplation, an inner calm and detachment, imbued with a deep God-centeredness, gives rise to a new awareness of self and others. Merton's insight is borne out in many saints in whom deep unity with Christ led to a higher plane of consciousness and transcendence in love. Francis of Assisi's sense of the world as the cloister of God and all creatures as his family is a good example of this new consciousness. In his *Canticle of Creatures* Francis expressed his harmony with creation, which reflected his deep inner unity with Christ in the cosmos. His life shows that the more one enters into the mystery of Christ and becomes reconciled within oneself, the more one lives at peace in the cosmos.[21] One begins to contemplate reality as a unity in God rather than fragmented individual parts. As Ken Wilber states, "When a person rediscovers that his deepest Nature is one with the All, he is relieved of the burdens of time, of anxiety, of worry; he is released from the chains of alienation and separate-self existence ... Thus, when one rediscovers the ultimate Wholeness, one transcends—but does not obliterate—every imaginable sort of boundary, and therefore transcends all types of battles."[22] Wilber describes the saints as "the growing tip of human consciousness because they discovered the higher levels of being through an expansion and precocious evolution of their own consciousness," thus transcending divisions within culture and society.[23]

Merton's transcultural consciousness reveals the evolution of consciousness that characterizes the second axial period. It is a complexified religious consciousness grounded in Christ, who is the integrating center of the human person and the cosmos. As Thompson notes, "a transcultural consciousness is capable of grasping the transcultural meaning of the risen Christ."[24] Merton wrote that, for a Christian, "a transcultural integration is eschatological ... It means a disintegration of the social and cultural self, the product of merely human history, and the reintegration of that self in Christ, in salvation history, in the mystery of redemption, in the Pentecostal 'new creation.'"[25] By transcending differences in union with Christ, one enters precisely into the difference of the other who is Christ; that is, a transcultural consciousness that grasps the transcultural meaning of Christ rises above the fragmentation of humanity by descending into solidarity with one's neighbor.

Merton's insights into the transcultural Christ reveal a mystic who has plumbed the depth of divine mystery to enter the heart of humanity and the heart of the world. The one who dies with Christ and makes the passage through the cross, letting go of everything that one holds onto for security, enters into the freedom of the Spirit of God, allowing God to be the God of one's life, thus rising with Christ into a new creation. Such a person has a new vision of the world that sees unity not difference, love not hate, family not separate tribes; such a person sees the world in the heart of Christ. Thus, as Thompson suggests, "belief in the risen Christ, rather than being an obstacle in the encounter of the world religions, becomes itself a means of furthering that very encounter."[26] To enter into the dialogue of world religions is not to transcend the humanity of Christ but rather to enter into the humanity of Christ as the symbol of all humanity, for the risen humanity of Christ is the power of God at work in the cosmos. Jesus Christ is the fully integrated person, and one who enters into Christ enters into the person of all persons in the heart of the universe. That is why, Thompson states, "Jesus is considered by Christians to be *the* revelation of God, for the finally integrated person is precisely the most complete way in which God could possibly reveal Godself to humankind in human terms ... Jesus draws us from that self-clinging which makes authentic relationships impossible."[27]

Merton's monastic approach to the fully integrated person, and Christ as the fullness of integration, is one of experience. To know God is to experience God; to love God is to experience God. The person who says he or she knows God and hates one's neighbor is a liar (1 John 4:20). In such a person, God does not exist. What

Merton indicates is that we must first come to know God within ourselves as our creator, sustainer, and lover. Only when we are at home with God within, believing in God's personal love for us, can we trust ourselves to grow in that love which challenges us to go beyond our comfortable boundaries. When we can let go and live in the poverty of being, free from those things that imprison us, then we can allow the gravity of divine love to pull us away from our small, finite selves to a higher plane of being where we find ourselves truly as we are, and in this true self, the freedom to love beyond constraint. We begin to see the world with a new vision and to love with a new love, as we move away from "absorption towards universality, from self-centeredness toward cosmification."[28] The path of transcultural consciousness opens up new insight to the mystery of Christ and provides a new Christian path in the second axial period.

If the risen Christ is the very center of unification in the cosmos, transcending cultural and religious differences, then the Christian who abandons Christ abandons any hope for unification and thus peace in the world; indeed, in the cosmos itself. The centrality of Christ is not an obstacle to the unity of world religions but the very source of this unity. Such an idea supports Christ as a coincidence of opposites whose fullness involves diversity and difference; the greater the opposites, the greater the fullness of Christ.[29] Perhaps if we did not use the title "Christ" and spoke instead of the fully integrated person at the heart of the cosmos we could grasp more readily the need to experience this person as the source of wholeness and healing within our own lives and the life of the world. *This* person is salvific. As Panikkar points out, to say "Jesus is the Christ" means that Jesus is not any person but *the* fully integrated person in whom God has revealed himself in the most complete way. Anyone who proclaims him- or herself "Christian" proclaims belief in the risen Christ and must be on the way toward development of a transcultural consciousness and, thus, transcultural encounters.

Bede Griffiths

The relationship between the fully integrated person and the risen Christ finds a clear visible expression in the life and works of Bede

Griffiths (1906-1993), a Benedictine monk who lived in India for about twenty-five years. Dom Bede Griffiths, born Alan Richard Griffiths, had a conventional English Edwardian childhood. At the age of seventeen he underwent a mystical experience of nature which was decisive in shaping the future course of his life. He attended Oxford, where his tutor was the renowned C. S. Lewis, who later became a close friend. Upon completing his studies at Oxford, he, following the theosophical teachings of Henry David Thoreau and Ralph Waldo Emerson, joined with two other young men in an attempt to live without the benefits of modern technology. This experimental lifestyle had a significant impact on Griffiths, who eventually converted to Roman Catholicism and soon after joined the Benedictine Order at Prinknash, taking the name of Bede. He accepted an invitation to establish a monastery in India, eventually taking over a monastic ashram in Kerala called Shantivanam, which had originally been founded by two French priests in the 1970s. This was to be the turning point of his life and the key to his life's work. Becoming a spiritual leader and writer, he remained at Shantivanam until the end of his life.[30]

Dom Bede went to India in the hope of finding there what was lacking not only in the Western world but in the Western church. In his book *The Marriage of East and West* he writes,

> We were living from one half of our soul, from the conscious, rational level and we needed to discover the other half, the unconscious, intuitive dimensions. I wanted to experience in my life the marriage of these two dimensions of human existence, the rational and intuitive, the conscious and unconscious, the masculine and feminine. I wanted to find the way to the marriage of East and West.[31]

Although he sought to amalgamate Christianity with Hinduism, Bede always remained a Christian monk. His devotion, however, was not to Christianity as an institution, still less to Roman Catholicism, but to Christ, whom he saw as the cosmic person. He expressed his dissatisfaction with the Western church, which, in his view, stifled the mystical element in Christianity—a sentiment expressed by Panikkar and others as well. Without relinquishing the centrality of Christ or the essential meaning of the church, Griffiths followed a path of contemplation that was deeply influenced by Hinduism. His experience as a Christian among Hindus opened up for him the depth of the mystery of Christ and the importance of interreligous dialogue as the source of fruitful growth for Christianity itself. In his book *The Marriage of East and West*, he urged the Western world to expand its horizons by entering into conversation and experience with the East. He writes,

> The Western world has to rediscover the power of the feminine, intuitive mind, which has largely shaped the cultures of Asia and Africa and of tribal people everywhere. This is a problem not only of the world as a whole but also of religion … All the Christian churches, Eastern and Western, have to turn to the religions of the East, to Hinduism, Buddhism, Taoism and the subtle blend of all these in Oriental culture … if they are to recover their balance and evolve an authentic form of religion which will answer to the needs of the modern world.[32]

Griffiths approached dialogue with other religions, in particular Hinduism, as a form of mystical experience. He indicated that in dialogue one takes the religion of the other as seriously as one's own. One does not have to step back from one's own belief in Christ as the Alpha and the Omega; rather, it is precisely this belief that is the starting point for entering into dialogue. His own experience led him to realize that "Christians so easily dismiss Hinduism as pantheism, monism or polytheism without realizing that, although those elements are present, God, the Word of God, is revealing Himself to them, and the Spirit of God is present to them."[33] He realized that conversion is the only way interreligious dialogue can bear fruit, since conversion involves a turning toward the other and letting the other be. Through conversion we acquire respect for the other and the realization that we have something to learn from them. Through our openness to the other's tradition and an honest search for truth, we can come to a better understanding of Christ and the mission of the church. He wrote, "Each religion has to purify itself and discover its own inmost depths and significance and then relate it to the inner depth of the other tradition. Perhaps, it will never be achieved in this world, but it is the one way in which we can advance towards that unity in truth, which is the ultimate goal of humankind."[34] Elsewhere he states, "It is only in the awakening of the contemplative spirit, of a transcendent consciousness, that we come to a vision of unity."[35] In an article entitled "Christianity and the Eastern Religions," Wayne Teasdale illuminates Bede's penetrating insight into interreligious dialogue by saying, "Doctrines (the domain of words and concepts) always occasion serious differences, misunderstanding and precious little agreement. But when Christians, Buddhists, and Hindus meet in silence and meditation, there is usually a deep experience of union and communion transcending the divisions that conceptualization brings. There is a discovery of that place of depth from which the various paths have arisen."[36] Prayer is the common language among world religions when it is spoken in silence and the meeting place where mutual understanding unfolds.

Although Bede traveled to India in the hope of filling up what was lacking in his Christian soul, he entered into a new religious experience that centered less on doctrine and more on mystery. Judson B. Trapnell, who wrote his doctoral dissertation on Bede's theory of religious symbols and practice of dialogue, describes him as a "cultural bridge" between East and West. He found that a chronological reading of Bede's writings reveals a close relationship between the stages of development in his mystical life and his position on the status of non-Christian religions. When Bede first moved to India, for example, he endorsed a "theology of fulfillment," according to which the numerous non-Christian religions are seen as providential means through which God prepared humanity for the revelation of Jesus Christ, a revelation that in turn fulfills all other religions. But after moving to Shantivanam and through the deepening of his insights about the divine mystery in meditation, he spoke of the "one spirit of all religion" and shifted to a theology of "complementarity." This latter theory suggests that each religion reveals a unique aspect of the one truth, aspects that when compared are not contradictory but complementary, like different colors within white light. Beneath this level of understanding of the uniqueness of each tradition, however, one may discover at "the deepest level" a "fundamental unity" or point of convergence of the various religions. This encounter takes place in the depths of contemplative experience, which is a way of saying that the deepest form of dialogue is not with words. Bede identified this point of convergence as "the cave of the heart," a mystical metaphor he borrowed from the Upanishads.[37] By living in this "cave of the heart," Bede found a deep inner source of unity, similar to what Merton described, that would lead him to the transcultural Christ. In his book *Return to the Center,* he reveals this deep unity.

> In meditation I can become aware of the ground of my being ... I can get beyond all these outer forms of things in time and space and discover the Ground from which they spring. I can know the Father, the Origin, the source, beyond being and non-being, the One "without a second."[38]

Does Bede suggest that truth is relative or that Christianity possesses only partial truth? One could infer such an idea if his Christology is not followed through completely. At one point in his journey, however, he realized that the divine mystery that underlies all mystery is not apprehended "by word or thought but by meditation on the Mystery." Truth is not of the mind, for the mystery of truth "is where all human reason fails." Truth is the experience of love. In Bede's view one does not so much possess truth as one is possessed by truth. "In each tradition," he writes, "the one divine Reality, the one eternal Truth, is present, but it is hidden under symbols ... Always the divine Mystery is hidden under a veil, but each revelation (or 'unveiling') unveils some aspect of the one Truth, or, if you like, the veil becomes thinner at a certain point."[39] In the life, death, and resurrection of Jesus, however, "the veil is pierced, the Mystery shines through." Bede realizes that neither word nor thought but meditation alone on the mystery can pierce the veil. He writes, "This is where all human reason fails. All these words, Brahman, Nirvana, Allah, Yahweh, Christ, are meaningless to those who cannot get beyond their reason and allow the divine Mystery to shine through its symbol. This is done by faith. Faith is the opening of the mind to the transcendent Reality, the awakening to the eternal Truth."[40] Although Bede spoke of Mystery underlying the world religions, he did not endorse a superficial syncretism in which each religion loses something of its own. Rather, he spoke of a clear commitment and a firm, clear faith, while at the same time being truly open to other ways of faith and commitment.

While there is a unity of experience among the religions there are also differences, and it is precisely the differences that enable a union of religious experience. Bede writes,

> I often use the illustration of the fingers and the palm of the hand. Buddhism, Hinduism, Islam, Judaism, and Christianity are all separate in one sense. But as you move toward the source in any tradition, the interrelatedness begins to grow. As one might say, we meet in the cave of the heart. When we arrive in the centre, we realize the underlying unity behind the traditions. But I'm suspicious of attempts to mix them on the outer level.[41]

Bede's approach to other religions was, we might say, an evolution of consciousness in Christ. He realized that this *mystery* of Christ "which lies at the heart of the gospels and the evolution of Christianity ... when known in its ultimate ground is one with the mystery of Brahman, Nirvana, Tao, Yahweh, Allah. It is the one Truth, the one Word."[42] As his life deepened in the East, it also deepened in the mystery of Christ that grew within him. Dialogue with Hinduism did not diminish Christ but conversely enhanced the mystery of Christ for Bede. At one point he wrote, "I would say honestly that to me there is a fullness and finality in Christ which I don't find in others. But I wouldn't press that on the Hindu obviously. I would rather simply emphasize the distinctive character of Christian revelation and of Christ."[43] It is interesting that Panikkar revealed a similar insight when he wrote, "There are Indian gurus of great and profound spirituality who tell me: 'The final stage in spiritual evolution is the discovery of the Christ.' They realize that [Jesus] Christ is the most sublime Epiphany that has ever existed on earth."[44] In one of his final statements on Christian mysticism Bede said, "In gathering all things, all matter into one in himself, he (Christ) transforms the world, bringing the cosmos, its matter and its processes, back to its source in the transcendent Reality which he calls 'Abba, Father!' This is unique."[45] Again shortly before he died Bede expressed his conviction of the centrality of Christ:

> In the historic revelation of God in Christ, Jesus experiences God (the Absolute), as Father, the ground and source of

Being, and as Spirit, that is as Self-giving Love, and himself as one with the Father and the Spirit. This seems to me to be the most profound revelation of the Mystery … This is at once the advaitic experience of non-duality—"The Father and I are one"—and the experience of personal love. Beyond the twilight of non-duality was the light of Trinitarian love.[46]

The profound insights of this monk reflect a person who was, in the words of Thomas Merton, fully integrated and thus who lived in the risen Christ. He described Christ as the "Self" of redeemed humankind, restoring it to its being in the Word. Christ, who is the Word of the Father, is the express image of God in which the plenitude of the Godhead is reflected; thus, each human being, each particular image, reflects this divine light according to its capacity. Bede claimed, "Because the Word in whom exists the plenitude of being is the center of Trinitarian love, we are plunged in the ocean (of divine love), we are immersed in it. We who are like drops of water are dissolved in it, but still we cannot comprehend it."[47] "The Word alone," Bede states, "comprehends the Father. "The Word is hidden and in that Word the Spirit is present, the Spirit which is in all creation and in the heart of every person, and in our own inmost being. The Spirit is Love, a love which penetrates every atom in the universe, which fills every living thing, which moves the heart of every person, which gathers all into unity. In that Spirit we are all one in the Word, each one unique in him or herself, reflecting the light of the Word, and in that Word we are one with the Father, the Source of all."[48] As we read in the Gospel of John, "That they may be one; even as You Father are in me, and I in You, that they also may be in us" (John 17:21).

Through years of contemplation and meditation on the mystery of Christ, Bede realized that "Christ is incarnate in every person, or rather he is incarnate in the whole universe. For as all are one Person and form the body of Christ, so the whole universe is one body, one organic whole, which comes to a head in the human person … Christ is the indwelling Spirit, the Self, of the universe, who redeems it from the dispersion in space and time, and unites all its diverse tendencies in one body in himself."[49] While Bede's words may sound pantheistic, a closer reading of his ideas shows that the particularity of the person of Jesus Christ leads to the universality of all things in Christ. In his view, each individual person has a unique character and is a unique image of God. As one grows in union in Christ in a deep personal union, one becomes more universal. Bede writes, "You do not diffuse yourself, you do not lose yourself as you grow in knowledge and understanding and love—you extend. You do not cease to be a person. You become more deeply personal. Jesus is the most personal being in history, and yet He is the most universal."[50]

Although Bede was not a scientist, he thought in evolutionary terms, much like Teilhard de Chardin. Although he opposed the materialism of modern science, he had a mystical view of the physically evolving cosmos. "Christ is the term of the evolution of the universe," he wrote, "which is evolving in space and time and which achieves its true being in him." In his book *Return to the Center*. he describes the evolutionary body of Christ as follows:

> The body of Christ was formed from the matter of the universe, from the actual particles, protons, electrons, neutrons, atoms, molecules, which constitute matter. Just as these particles are organized into living cells and begin to perform the functions of life; and just as the living cells are formed into animal organisms and begin to perform the functions of animal life; and as the animal organism with its atoms and molecules and cells is formed into a human being and beings to exercise rational activity; so too in the resurrection these same particles of matter, these living cells, this animal organism, were formed into a 'spiritual body,' a body filled with divine life and participating in the divine consciousness.[51]

In a manner similar to Teilhard's process of anthropogenesis, noogenesis and Christogenesis, Bede held that matter is unified in human consciousness, and human consciousness is unified in the divine. The evolution of matter from the beginning leads to the evolution of consciousness in humans; it is the universe itself that becomes conscious in the human person. "It was the whole of this organic process of evolution from matter to life and consciousness in man that Christ assumed into himself."[52] Bede says that this is the full meaning of the incarnation—the assumption of the whole universe and the whole of humanity into the divine life. The evolution of the universe is from matter to life, to consciousness in the human, and from human consciousness to divine life and consciousness. He concedes that the movement toward the fullness of consciousness is revealed in the resurrection.[53] The resurrection is the point of emergence of this divine life in the human person, a divine life that had been present from the beginning through the Spirit, latent in matter from the beginning. As Bede says, "Christ opened up the depths of the unconscious to divine consciousness. He redeemed the whole creation by opening it to the divine life, the life of the Word, which filled his human consciousness." The whole creation has to pass into human consciousness, to become invisible in the human person, and we ourselves have to pass into our self, to transcend ourselves, and become one with that eternal truth.[54] This progression toward the fullness of consciousness takes place against the background of sin. Bede says that we bear the sin of Adam in us, but, in Christ, our human consciousness has been opened to the divine. He quotes the following from the Muslim tradition: "Your own existence is your greatest sin—that is, your separate existence apart from the Self."[55] As we awake to the consciousness of sin, so we wake to the consciousness of redemption. The struggle against sin is the struggle for a higher consciousness in Christ. Bede writes,

> Christ opened up the depths of the unconscious to divine consciousness. He redeemed the whole creation by opening it to the divine life, the life of the Word, which filled his human consciousness. It is said that every man recapitulates in the womb not only all the stages of the evolution of matter but also all the stages of the evolution of human consciousness ... He [Christ] took upon himself the sin and suffering of the whole world; he recapitulated all its stages and brought it up into the consciousness of the Word. This is the meaning of redemption. Each of us was "in Christ" on the cross, just as each of us was in Adam when he sinned. We bear the sin of Adam in us—the sin of man—but in Christ our human consciousness has been opened to the divine. For most of us this still lies buried in the unconscious—both our sin and our redemption—but as we wake to the consciousness of sin, so we wake to the consciousness of redemption. We are fully redeemed only when our human consciousness, with all that was formerly buried in the unconscious, is possessed by

the divine consciousness, and we know ourselves and humanity and the universe in the light of the Word. [56]

Bede is not saying that redemption is a matter of consciousness, as if sin does not really affect the human person. Rather, it is the reality of sin that we must become conscious of; redemption is recognizing our sinfulness and the healing of our divisions in and through Christ crucified. Bede also indicated, however, that "each of us has to work out his or her salvation within the pattern of the whole. No one sins in isolation and no one is glorified in isolation."[57] It is only when we see the sufferings of the world, the inequalities of the world, the injustices of the world within this pattern of the whole, in their relation to the eternal order, "in Christ," that we see their meaning and their justification.[58] Our destiny, he states, is to be one with God in a unity that transcends all distinctions, and yet in which each individual being is found in his integral wholeness. "It is only in the awakening of the contemplative spirit, of a transcendent consciousness, that we come to a vision of unity."[59]

The path to unity for Bede is love. Love is the ultimate mystery of being and the ultimate truth. Love is the essential structure of reality, the metaphysical basis of all that exists, the eternal pattern of the universe. Christ is the center of all reality because Christ is the fullness of love, that love which is the most complete integration and union of divinity and humanity. In his *Return to the Center*, Bede describes a metaphysics of love that is in harmony with the Christian mystical tradition and encapsulates the insights of Teilhard and Panikkar. He writes,

The Father gives himself in love to the Son, who is the very form, the expression, of his love, and this love returns to the Father in the Holy Spirit, who unites Father and Son in the eternal embrace of love. Love giving itself, losing itself and finding itself in love, and Love returning to itself, giving itself back in love—this is the eternal pattern of the universe. Every creature in the depth of its being is a desire, a longing for Love, and is drawn by Love to give itself in love. This is its coming into being, this response to the drawing of Love. At the same time it is being continually drawn to give itself in love, to surrender to the attraction of Love, and so the rhythm of the universe is created. The nucleus throws out its protons and electrons and they circle round it, held by the attraction of Love. The sun throws out its planets and they circle round it, held by the same attraction. The cell divides and then again unites, building up the body in love. It is the same with sexual love. The male is drawn to give himself in love to the female, and the female is drawn to give herself back in love to the male. There is a continual dance of love, a continual going and returning. Ultimately it is the one Love giving itself continually so as to create this form and that form, building up the universe of stars and atoms and living cells, and then drawing everything back to itself; everything coming into being in the Word as an expression of love, and everything returning to the Father, to the Source, in the love of the Spirit.[60]

Bede, like Teilhard, saw love as the very nature and structure of reality because the cosmos, created through the divine Word, is a perichoresis of trinitarian love centered in Christ. As Bonaventure and Scotus remind us, Christ is first in God's intention to love and to create. One who enters into this love at the center of one's soul, enters into this love at the heart of the universe. The evolution of human consciousness toward transcultural consciousness (and complexification of the Christ mystery) is a deepening of oneself in the risen Christ in whom the power of love transcends differences and thus unites.

The import of Bede Griffiths's works lies in the capacity of the human person for divine life, for the fullness of love and the unity of all things in love. The mystery of Christ is a mystery of unity in love, and only by becoming oneself in Christ does the unifying love of Christ become reality in the world. Dom Bede found this love and thus found the center of the universe within his soul. Lawrence Freeman claims that Bede can be of tremendous help to today's Christian because he shows "that it is possible to follow a mystical path and remain within the institutional church. People who feel that there must be one God, one reality, behind all religions, that God did not create a divided humanity, have found in Bede that their instincts are clarified, developed, most importantly,

lived."[61] We might describe Bede Griffiths in the same way that
Ewert Cousins describes Raimon Panikkar, a "mutational person,"
a totally integrated being who has brought insight to our human
evolution of consciousness. He has given us a wisdom Christology for our time, a rediscovery of the Christ-event in the context
of the larger, dynamic, and interrelated world of reality—a renewal
of the Pauline theme of the body of Christ: "For in him the whole
fullness of God dwells bodily, and you have come to fullness of
life in him, who is the head of all rule and authority" (Col 2:9-
10). Bede reminds us that the life of Christ is our life; the mystery
of Christ is our mystery. We must enter into it, make it our own,
discover ourselves in it, and make it alive.

CHAPTER SEVEN

CHRISTOLOGY REBORN

The works of Teilhard, Panikkar, Merton, and Griffiths provide roadmaps to the mystery of Christ in an evolutionary universe and a humanity marked by a new level of consciousness. As we move from first-axial-period consciousness into the second axial period, we find ourselves related not only vertically to God but horizontally to the global tribe of humanity and the earth as well. This new age of consciousness is one that includes not only freedom, autonomy, and self-transcendence but also relatedness. The new consciousness of the second axial period impels us to search for new ways of relating to God incarnate in the cosmos by contemplating the divine in our midst. We are in a new world today beyond what the fathers of Chalcedon could possibly have conceived. Teilhard claimed that Christianity is reaching the end of one of the natural cycles of its existence. "Christ must be born again," he said; "he must be reincarnated in a world that has become too different from that in which he lived."[1]

It is indeed this "new birth" of Christ that confronts us through the insights of our mystical writers. All four writers had contact with Eastern religions and it is interesting to note their common concern: Christology must detach itself from its Western, intellectual form. The word "detachment," however, may not properly describe the convergence taking place. In an evolutionary world, simple isolated structures must let go (or die) for the sake of more complex unions; so too, Christology as a Western (and hence isolated) doctrine is undergoing a transformation, a complexification, through the evolution of human (and Christian) consciousness. The complexification of Christology in the second axial period marked by global consciousness seeks a new way to do Christology, that is, to reflect on Christ in view of complexified religious consciousness. What is taking place, therefore, is not really a break from a Western intellectual form of Christology but an evolution of this form through global consciousness. While historical consciousness is integral to Christology today, it is not entirely sufficient to plumb the mystery of Christ. The works of Panikkar, Merton, and Griffiths suggest that global consciousness is yielding to a more complexified consciousness of Christ. It is a new consciousness of relatedness that requires prayer, contemplation, dialogue, and scripture as well as critical reflection. While there is a necessary role for the intellect in the study of Christ, "doing Christology" requires participation in the mystery. Spirituality, I maintain, is the heart of Christology in this new millennium. It is noteworthy that three out of the four mystics we have discussed are from religious traditions that emphasize the humanity of Christ—for Teilhard, the Ignation tradition, and for Merton and Griffiths, the monastic Benedictine tradition. The human reality of Jesus is the most focused statement of what we are about with God and what God is about with the world. It is through reflection on the significance of Jesus' humanity that our four writers reveal a new "vernacular" theology, a way of doing theology "from below" that emphasizes experience as the basis of intellectual reflection and insight.

The term "vernacular theology" is not new but finds its roots in the Middle Ages. Although Western Christology since the Middle Ages favored a scholastic approach to understanding Christ, Bernard McGinn has shown that scholasticism was not the only means of doing theology in the Middle Ages. Monastic theology predominated up to the rise of scholasticism, and this approach combined reflection on the experience of God with scripture. McGinn describes a third type of theology in the Midde Ages (alongside that of monastic and scholastic theology) that emerged among holy men and women. He calls this type "vernacular" theology and distinguishes it by linguistic expression in the vernacular tongue. It is a type of grassroots theology arising from the experience of God whereby the authority to teach comes about not *ex officio* but *ex beneficio*, or by the gift of grace.[2] The importance of vernacular theology, which arises from the inner experience of God, is that reflection on God is integrally related to Christian life; indeed it arises out of Christian life. That is, the desire for God leads to the search for God and hence to love of God. Vernacular theology is essentially reflection on the love of God and the God who is love. It is not inimical to intellectual tools of analysis; it simply finds them irrelevant, especially since vernacular theologians of the Middle Ages were unlearned. Vernacular theology, however, does look to scripture and tradition as the primary sources for apprehending the mystery of God.

The mystical insights of Teilhard, Panikkar, Merton, and Griffiths have ushered in a new type of vernacular theology in the twenty-first century. It is no longer a theology of the unlearned but of global experience, arising from the spiritual search for Christ amidst evolution, religious plurality, and difference. Vernacular theology today is a way of going about the world; it is not a matter of analyzing concepts but of *doing* theology, shifting the context of theology from the rigor of academic discipline to the context of life and holiness. It is the theology of the poor person— not necessarily the economically poor, but the one who recognizes his or her need and dependency on God and neighbor and lives in openness, receptivity, and gratefulness. Vernacular theology is engaged theology whose primary approach is through historical consciousness, spirituality, and dialogue. We see a type of vernacular theology in the rise of feminist, Asian, African, and Hispanic Christologies, which indicates that Christology is no longer a universal, European-centered understanding of Christ but a reflection on the local experience of God incarnate, which arises from the context in which one lives or one's experience of the world. Vernacular theology in the second axial period is a reflection on Christ in view of evolution and is not content to confine itself to philosophical analysis (although it does not deny such analysis). Rather, it is doing theology by way of spirituality. Thus, it involves the inner dimension of a person where ultimate reality is experienced and out of which one's actions flow. Spirituality, or the "doing of Christ" in the world today, must be the source of understanding Christ in this new age.

The importance of a vernacular theology of global consciousness is that it opens us up to the mystery of Christ in a way that a more philosophical, intellectual approach cannot. It is a renewal of the Word incarnate as a spoken event, the *kenosis* of God's self-giving love in creation and receptivity to that love—a living and dynamic Word that resonates throughout creation. The dynamism of the divine Word spoken throughout creation allows us to consider the Christ-event from its inception in history not as a static event but as an evolutionary one. Jesus brings newness in himself, a "new creation." The newness of Jesus initiates a continuing drama and dialectic in the history of the world; human life and history take on a new theological intensity. Teilhard perceived that the Christ-event brings a new directionality into evolution, which begins with the human person. God enters into the evolutionary process in Jesus Christ, initiating a new incarnational dynamic that moves laterally rather than upward. The historical event of Christ is a divine "Big Bang" in the history of the universe. This new power generates a new history. Christ brings a "new heart" to humanity, both on the individual and on the collective plane. As Teilhard points out, humanity becomes a new "creative center" within the evolutionary process in such a way that the path of evolution changes. The evolution "from cosmogenesis to Christogenesis" now takes place under the influence of a divine creative power that has been infused into humanity. Christ becomes the concrete focal point or Omega of a new centration, or, in Teilhard's language, socialization or planetization.

Christology can no longer be a reflection on how Jesus Christ is truly God and truly human, but now we must also reflect on how Christ is truly cosmic. We must take seriously Teilhard's cosmic nature of Christ if evolution of the cosmos is aimed toward the fullness of life in God. In Teilhard's view, the predominant concern of theology in the first centuries of the church was to determine, intellectually and mystically, Christ's relation to the Trinity. In our own time, it is to define the link between Christ and the universe; how they stand in relation to each other and how they influence each other. In the life, death, and resurrection of Jesus, the universe reaches the climax of its history of self-transcendence toward God. His bodiliness, like ours, is essential to the human person. It connects us with the whole material world. Through our bodies the whole world belongs to us and we are part of the whole. The reality of Jesus signifies to us that the whole cosmos is on a journey to God in and through the human person who is the growing tip of the evolutionary movement. Because the body of Jesus emerges from the evolutionary process in the same way as our bodies do, and because matter itself has a spiritual potency signified by the incarnation, it is reasonable to speak of a "third," or cosmic, nature of Christ, if we understand nature to be not a "thing in itself" but an outcome of influences and relationships within the evolutionary process. As the risen Word incarnate, Christ indeed is related to the whole cosmos, and the whole cosmos finds its meaning in Christ.

It is in view of this integrated relationship between Jesus and the cosmos that Karl Rahner spoke of Jesus as the symbol of God. Paul Knitter summarizes Rahner's "transcendental Christology" by saying,

> The essential Christian claims that Jesus is divine—one person with both a divine and a human nature—cannot be a mystery that we must accept on external authority or naked faith. Although it will always be beyond full comprehension, it must make sense on the basis of our experience of ourselves and our world; it "must possess for man a genuine intelligibility and desirability" ... What happened in Jesus of Nazareth represents the fulfillment of what we are as human beings.[3]

Jesus is the symbol of what human beings really are and what is intended for all creation. As symbol, he provides the assurance that the infinite hopes and strivings of our human nature can be realized: God and the world can be one in unity. As we noted in chapter 5, the word "symbol" means "to throw together." A symbol connotes union and conjunction and thus mediates identity. When we say that Jesus is the symbol of God we are saying that God's self-utterance, his *ex*-istence is realized in the humanity of Jesus. Since the incarnation fulfills the potential of every human person, the incarnation symbolizes the union of God and humanity/creation for every person, whether or not they explicitly know Christ. To say that Christ is the symbol of God is to say that Christ is the Word of God expressed or projected outward in space, time, and history. This expression/projection of the symbol of the Word discloses and manifests what is present. The symbol makes real and reveals that which is otherwise hidden or concealed, that is, the centrality of divine life. The symbol of Christ functions as the meaning of God for us and discloses to us the true meaning of our lives in God. Christ is symbol of the trinitarian mystery of God by which God is able to communicate God's life completely to another and to be united with the other as beloved in a perichoresis of love.[4] The symbol of Christ as symbol of God helps us to understand Jesus Christ as the one in whom all humanity finds its permanent openness to God. As the crucified and risen one, Jesus is the "anointed one," the one "filled with the Holy Spirit"; he is by nature what we are to become by grace, the eschatological fulfillment of God's plan of creation. Jesus is not the great exception to humanity but its culmination and fulfillment. In Jesus we experience the fact that the mystery of the human person, which is not for us to control and "which is bound up with the absurdity of guilt and death is, nevertheless, hidden in the love of God."[5] Jesus the man is truly God and the true divinization of our humanity; in him we attain full humanization.[6] Since the life, death, and resurrection of Jesus is the symbol for all human life, the meaning of Christ cannot be tribal or contained in one particular group. Rather, the meaning of Christ is the divinization of the whole cosmos. That is why Christ is the goal of an evolutionary universe and of human life in evolution. As Panikkar states, "Christ is the symbol of the whole of reality."[7] In this way, Panikkar notes, Christ's reality is not exhausted with Jesus' historicity.[8] Jesus is the Christ, but Christ is more than Jesus because "Christ is that central symbol that incorporates the whole of reality."[9]

While Christ symbolizes the whole of reality, the whole of reality is intended to symbolize Christ. This is the essential meaning of the primacy of Christ and the "christic universe" described by Teilhard. The potential for God in creation is realized in Jesus Christ, but that potential has yet to be completed in the fullness of Christ which lies in the capacity of every human person to image (and thus express) Christ. How do we understand this "congruency" between the human person and the fullness of Christ? Rahner himself posed the question in this way, "If the human life of Jesus is God's *ex*-istence, that is, God's self-communication, so that human reality is God's reality and vice versa, then what do our lives mean when they are first and last the life of God?"[10] Rahner's question is at the heart of Bonaventure's thought as well. Bonaventure developed a theology of congruency between the incarnate Word and the human person in such a way that the reality of Christ finds its form in humanity and creation. Without the expression of Christ in humanity and creation, Christ does not exist in creation. Or we may say, Christ exists precisely because humanity and creation have been united together in union with God through the resurrection of Jesus. Because Jesus is the symbol of our humanity, Christ is not *a* person but *every* person in whom the power of God is expressed. This is the essential meaning of Panikkar's christophany. The human person can express God because the human person is the image of God and has the capacity for God (*capax Dei*). Just as Jesus was raised from the dead by the power of God to become the Christ, the fully integrated person, we too must rise from the dead. Unless the power of God is at work in us raising us to new life— spiritually and ultimately physically—the fullness of Christ is not complete in the universe. To say that Jesus is the symbol of God is to say that Jesus' life is our life, his destination is our destination. Christ is the most focused statement of what we are about in the universe.

In translating Bonaventure's *Soul's Journey into God*, Ewert Cousins noted the peculiar structure of Christology in this work; the mystery of Christ is a mystery of opposites: divine and human, eternal and temporal, Alpha and Omega, first and last.[11] What is striking about Bonaventure's coincidence of opposites, as it marks the mystery of Christ, is that it is described symbolically. The coincidence of opposites is not a doctrine of Christ; rather, it describes the living reality of this mystery. One cannot begin to know Christ, Bonaventure indicates, without discipline, grace, and a humble desire for God. The mystery of Christ cannot be studied or formulated; rather, it is a discovery through prayer, grace, virtue, illumination, the search for truth, self-knowledge, and knowledge of God. Bonaventure describes Christ, the coincidence of opposites, at the penultimate stage of the journey to God. He uses the symbol of the temple to describe the soul and the symbol of the Mercy Seat to describe Christ, the coincidence of opposites. He writes,

> Look at the Mercy Seat and wonder
> That in him there is joined
> The first principle with the last,
> God with man, who was formed on the sixth day;
> The eternal is joined with temporal man,
> Born of the Virgin in the fullness of time,
> The most simple with the most composite,
> The most actual with the one who suffered supremely
>   and died,
> The most perfect and immense with the lowly,
> The supreme and all-inclusive one
> With a composite individual distinct from others,
> That is, the man Jesus Christ.[12]

If the symbol mediates one's identity and thus functions as an expression of the whole, it is interesting that Bonaventure describes the symbol of Christ and the union of opposites in the context of contemplation. The call to "look" at the Mercy Seat evokes the idea of "gaze,"[13] which to the medieval mind involves "touch," "taste," "experiment" as well as "see." Contemplation is an engagement with the other, a movement out of oneself into the other by way of vision, and leading to union. Before one contemplates Christ, one must enter into oneself and discover oneself as image of God through prayer, the practice of virtues, use of the spiritual senses, and divine light—all of which is made possible by a sincere love of Christ.[14] This too is the insight of Panikkar, Merton, and Griffiths. That is, we cannot apprehend the meaning of Christ for us unless we are prepared to receive that meaning within us.

The symbol of Christ is an invitation to union with God through contemplation, a penetrating gaze within ourselves and on Christ crucified, a gaze that accepts the disclosure of God in fragile human flesh. This encounter is self-revelatory because the God who comes to us in the person of Jesus Christ is the ground of our being as well. To accept God in our own human lives, symbolized by Christ, is to accept the ground of our identity. In the symbol of Christ, God reveals himself to us, and we are revealed to ourselves as we begin to see the truth of who we are—our image. The more we contemplate Christ, the more we discover our identity; and the more we discover our identity, the more opposites within us are reconciled. Thus, Christ comes alive within us. Contemplation is creative because it transforms the one who gazes in the mirror of the cross into a reflection of the image itself. That is, the more we contemplate Christ, the more we discover and come to resemble the image of God in whom we are created. This image of God, brought to light in the one who gazes upon the symbol of the Mercy Seat (Christ crucified), is the basis of transformation, or "birth" of Christ in the believer. As we come to be who we are called to be in relation to God (self-identity), God shows himself to the universe through his constant and continual creation of the self. The self that comes to be through a union with God in love is the self in which God is reflected, that is, the image of God. Christ is the symbol by which the image in which we are created is brought to perfection, that is, the fullness of life in God.

We can relate Bonaventure's symbolic Christology to Merton's "integrated person" as one in whom opposites are reconciled or, we might say, as a unifier of opposites. Merton's hope for a new planetary consciousness led him to this realization: "If we want to bring together what is divided, we can not do so by imposing one division upon the other or absorbing one division into the other ... We must contain all divided worlds in ourselves and transcend them in Christ."[15] That is, we must come to a new level of transcultural consciousness by which we overcome all divisions within ourselves and thus come to realize what divides us from others. For Merton, as for Bonaventure, only a life-giving, prayerful relationship with God in union with Christ, who is the symbol of the fully integrated and unified person, can one attain "inner calm and personal and cultural detachment."[16] The transcultural person "does not repudiate society but lives and acts within it from a standpoint higher than any limited society can offer," or, as Merton writes, "he has embraced all of life." Such a person is "guided not just by will and reason but by spontaneous behavior subject to dynamic insight."[17]

Bonaventure, like Merton and others, helps us realize that the true meaning of Christ cannot be confined to confessional formulas; Christ must be contemplated. Spirituality is the key to Christology, especially in the second axial period; and only one who is prepared to enter into the mystery of Christ can truly know Christ and live Christ as a co-creator of the universe. Life in Christ is born of the Spirit. It requires openness to the divine energies permeating our lives and our world. If the spiritual journey is the way into the mystery of Christ, it is love that forges an identity with Christ. As George Maloney states, "knowledge of another person leading to a person-to-person relationship in love is not merely a question of conceptual knowledge but existential—not 'what' but 'who' one is in the being encountered."[18] This idea underlies Panikkar's distinction between identity and identification as well. The individuality of the unique person can never be captured completely in a mental concept by physical description alone. The inner core of a person remains sealed off to others unless a mutual love engenders a deeper insight into *who* this person is. In our encounter with Christ today it is not enough to contemplate him as the historical person revealed in the Gospels. Nor is it enough to contemplate him in the encounter of faith through grace by which we become united with him and other members of his mystical body. We must also be able to meet Christ in the divine, continual act of creation, redemption, and sanctification of the total universe. In him we discover the Absolute, the beginning and end of all unity in the cosmos. Faith gives us the eyes to see not *what* God is but *who* God is. As Teilhard indicated, to foster a steady growth of faith, a divine milieu must be created in which Christ can be personally contacted. He is to be found at the root, the ground, of our very existence.[19] To discover Christ as the ground of our existence is to discover God enfleshed in our lives. Such discovery is not through intellectual rigor but spiritual depth, for only the Spirit of God can search the depth of God (1 Cor 2:10). Prayer that leads to contemplation prepares the human heart to enter Christ through the door of love, which, as the mystics claim, goes further than knowledge. Knowledge may lead us to the doorstep of God, but it is love that enters the mystery of God.

If the theologian of the second axial period is to search the depths of God then we need a "mystically informed theology."[20] Contemplation must be at the heart of theology. Bonaventure developed a method of theology early on in his career that combined the symbolic and the scholastic. It is the action of "uncovering, searching out, penetrating, or fathoming," and thus is the most appropriate theological method for allowing the depth of the mystery to unveil itself without destroying it.[21] Like the "seeker of pearls," the theologian sent by the Holy Spirit strives to descend to the bottom of the Christian mystery; but the penetration of the mystery is also an inquiry or examination, a method of reasoning. The theologian is one who brings to light the hidden depths of reality.[22] Bonaventure's method of searching the depths of things is similar to the gaze described by the postmodern philosopher Jean Luc Marion. To talk about the "depth" of some phenomenon is also to talk about its hiddenness at a depth that paradoxically reveals itself only in not showing itself.[23] Marion writes, "Depth does not indicate that 'behind' the phenomenon something else would be waiting to appear, but that the very appearing of the phenomenon—as a way (of Being) and therefore as a nonbeing—reveals a depth. The depth does not dub or betray the phenomenon; it reveals it to itself."[24] Similarly, for Bonaventure, the depth of theology as it flows from the mouth of the triune God reveals the mystery of God that is concealed in the depths of creation. The very act of creation discloses the depth of divine mystery that is concealed in the act of being revealed. Bonaventure's "method" redirects the novelty of its approach toward the internal disposition of the subject which alone can explain it. The searcher of divine depths must be on the journey to God. Indeed, only one filled with the Spirit can search the depths of God. Theology is to make God manifest so as to orient one toward an encounter through love between God (and God's act of revelation) and the human person in one's affective power. Love therefore becomes a conceptual determination at the junction of theory and practice. Emmanuel Falque writes, "any strictly theo-logical truth, one that has its roots in God, will no longer be content with its unique objective determination." "Such a truth," he states, "will take on a performative sense, one that is transforming for the subject that states it or it will not exist ... Knowledge through love is the only thing that puts in motion whoever comes to know them."[25] Spirituality, therefore, is the key to penetrating the depth of the divine mystery, and love is at the junction of theory and practice. God who is the source of all theology is a divine depth to be searched and fathomed by one imbued with the Holy Spirit.

In his *Soul's Journey into God* Bonaventure shapes the theologian in such a way that the development of spirituality in one's life corresponds to the depth of theological insight. One who attains the heights of contemplation, infused with the grace of the Holy Spirit, attains the height of theology, which is the fullness of revelation in the crucified Christ. He points out that theology, knowledge of God, cannot be reduced to intellectual thought alone. Theology does not serve the desires of rationality; rather, reason is feeble where the light of God is most brilliant. Bonaventure helps us realize that when theology is forced to obey the principles of rationality alone it ceases to be the "*logos* of God." True theology is not a doctrine but a living reality, revealed in a living person, the person of Jesus Christ. The key to knowledge of God revealed in Christ is contemplation, which is the fruit of prayer and the grace of the Holy Spirit. Contemplation is born from the "cave of the heart"; it is the vision of a heart centered in God by which one sees the depths of things in their true reality. The highest form of contemplation is not intellectual knowledge but knowledge through love—wisdom—and the highest wisdom is revealed in the crucified Christ. Bonaventure maintained that God, who is a Trinity of incomprehensible love, reveals that love in the mystery of the cross. Christ crucified is the mystery hidden from all eternity (Eph 3:9), and only one filled with power of the Holy Spirit, the fire of love, can enter into this mystery. As we journey into the heart of the crucified, the knowledge of God that draws us beyond ourselves impels us to leave behind the logic of the intellect and to rely on desire, the fire of the Spirit within; here the mind gives way to the heart and we are drawn to the one whom we can never fully understand but whom we desire from the depths of our being. In his *Itinerarium* Bonaventure writes,

> But if you wish to know how these things come about, ask for grace not instruction, desire not understanding, the groaning of prayer not diligent reading, the Spouse not the teaching, God not man, darkness not clarity, not light but the fire that totally inflames and carries us into God by ecstatic unctions and burning affections. This fire is God, and his furnace is in Jerusalem; and Christ enkindles it in the heat of his burning passion.[26]

Bonaventure speaks of this union with the crucified Christ as a "mystical death" brought about through the grace of the Holy Spirit. It is a "dying into love," being drawn into the mystery of the Father through the power of the Spirit's love flowing from the passion of the crucified Christ. The incarnate and crucified Word who expresses the Father's power, wisdom and goodness in creation becomes the manifestation of his overflowing love on the cross. At this highest stage of mystical union with the crucified Christ theology is perfected, for the knowledge of God becomes the experience of God as one passes over into the silent, hidden depths of the Father through the love of the Spirit and the humanity of Christ. This is the highest stage of wisdom, according to Bonaventure. One must be on the spiritual journey to God to know the wisdom of God expressed in the depth of the cross. Only through union with Christ and empowerment by the Spirit can one attain this wisdom. That is, one cannot *know* God in himself but as he *reveals* himself, that is, one must *experience* God, "for the Trinity is *substance* to a greater degree than our intelligence can perceive."[27] The human response to God cannot simply be one of intellectual contemplation but must lead the person into a union of love with the primordial mystery of love, that is, to be led back to the loving relationship between Father and Son. One who is inflamed by the Spirit of love passes over in Christ to the silent, hidden depths of the Father who is the fountain fullness of love.

Theology, the logic of God, is the logic of self-involvement and involves a kenosis of one's self. One who seeks God must be transformed into the love of the crucified in order to enter into what is disclosed as lasting and true. True knowledge is bound up with death, and unless one is willing to "die to the ego" in order to attain union with the other, truth remains elusive. For Bonaventure, truth is not a given but must be arrived at; it is not a "thing" but a disclosure. True knowledge is attained through union in love, and without love there is no truth and hence no real grasp of true reality. The Christian (and theologian) is one who, filled with the Spirit, contemplates the depths of the mystery, the fountain fullness of God's love revealed in the cross and returns love for love. Bonaventure writes, "Christ goes away when the mind attempts to behold this wisdom through intellectual eyes; since it is not the intellect that can go in there, but the heart."[28] We must let go of possessing knowledge and let ourselves be possessed by the God of overflowing love as that love flows into the heart of creation.

It is in light of a mystically informed theology arising out of a depth of contemplative insight that Panikkar and others warrant against any type of "post-Christianity." A post-Christian stance can arise only out of an ill-informed understanding of Christ that mistakes the centrality of Christ as the personalizing center of the universe for a white, Western, male Jesus, who was really Middle Eastern and Jewish. It is a rejection of the mystical Christ based on a rationalization of Christ. Bonaventure's theology of the Word speaks to Panikkar's concern by offering a Christology that is cosmic and hence inclusive. Christ is not merely *a* person but *the* person, the cosmic personal center of a God-directed universe. There can be no "beyond Christ" because Christ alone is the fullness of what we hope for in God. The source of this fullness is the Holy

Spirit, which Bonaventure presumes in his doctrine. Since Christ is the Trinity incarnate there can be no evolution of Christ without the Spirit. Yet Bonaventure never really elaborates on the role of the Spirit in the same way he elaborates on the role of Christ. His mature cosmic Christology overshadows the essential role of the Spirit. Although one could read a type of pan-christism into his doctrine, Bonaventure is fully trinitarian and aware that without the Spirit there is no Christ.[29] While some post-Christians today espouse a spirituality purified of Christ, as if the Spirit can act independently of the Father and the Son, such spirituality is bereft of any real theology. To dismiss Christ as the life of the world is to abandon any hope for the world in God. The Spirit is creative and unifying love precisely because the Spirit is the mutual love breathed forth by the Father and the Son. A spirituality of "Spirit" alone is a denial of the Trinity and hence a denial of Christ. Such denial can lead to a worldly immanentism that yields to privatized spirituality and fragmentation, since there is no real ground for relationships in creation. A post-Christian position that dismisses Christ is one without a transcendent, unifying center. Without a center it can wind up in a new type of hegemony, imperialism, or tribal religion because the very hope of unity in the world (which many post-Christians espouse) has no real basis of unity.

The renewal of Christ as the center of an evolutionary universe in the second axial period means an openness to the evolution of consciousness on a global scale and a new level of complexified religious consciousness, which brings a new understanding of Christ and ourselves in relation to Christ.[30] We must be open to a new way of doing Christology in this new age, not with the tools of philosophical analysis alone but with the searching of the Spirit, the contemplative heart, and the eye of wisdom. As we engage in the evolution of Christ today, we find that Christology is self-involving because the one who searches the depths of Christ must first discover the depths of Christ in one's own life. Such a discovery is the basis of our vocation in this universe as co-creators in Christ, the potential within us to be integrating centers in the universe. We must discover Christ as new life within us; we must allow ourselves to be changed to a new level of consciousness, a new level of transcendent love, and a new vision of the world. The vernacular theologians of our time, giving witness to the resurrection of

Jesus Christ, proclaim the evolution of human consciousness. They assure us that Jesus is no longer in the tomb but has risen, and we too have risen. The risen Christ is the inner power of this evolutionary universe that impels us to go forward into a greater unity in love despite the forces of separation. To be alive in God is to be alive in Christ. So why do we continue to seek the living among the dead? Christ is in evolution because we are in evolution, and the hope of a new creation lies within us.

CHAPTER EIGHT

CO-CREATORS IN CHRIST

If Christ is what we are about, then "doing Christ" is what we are created for, making the power of God alive in the universe through our lives. The idea of a God-centered life as the life of Christ in creation corresponds to Teilhard's notion of "co-creator." For God, "to create" is to unite Godself to God's work, that is, to involve Godself in the world by incarnation. God evolves the universe and brings it to its completion through the instrumentality of human beings. The human person is called to be a co-creator—a cooperator with God in the transformation of the universe. Thus, it matters what a human person does and how a person lives in relation to God, for only through his or her actions can she or he encounter God.[1] Theologian Philip Hefner describes the human co-creator in light of evolution:

> We are creators, by our very nature and experience. If we attend carefully to our experience, however, we recognize that we are as much created as creator. The scientific story tells that the processes of nature have created us, by means of evolution. We did not give ourselves our physical-chemical-biological composition, nor did we give ourselves brains and the culture that they make possible and necessary. When we turn to our religious traditions, we will see that they speak of our being created by God. These natural processes that have engendered us are declared to be the instrumentality of a divine creator ... There is a linkage, however, between the source of our being created and our own creativity. To the degree that evolving nature has created us, our own creating is taken up into that nature, so that we are nature's own creators, co-creators with the evolutionary process that has engendered us.[2]

138

While it might seem fantastic that we are co-creators in the universe, the meaning of being image of God (*imago Dei*) means the capacity to give birth to God in creation, that is, to "imitate" God in our lives. Bonaventure's understanding of the human person is one of co-creation in Christ. The incarnation of the divine Word in human flesh means that human nature is capable of being actuated by personal union with the divine; the incarnation reveals the deepest truth about the human person and his or her relation to the divine.[3] Teilhard captured this idea in his own way by conceiving the whole of natural evolution as coming under the influence of Christ, the physical center of the universe, through the free cooperation of human beings.[4] Christianity, he said, is intended to be a new "phylum of salvation that spreads its inner life to the rest of the universe, a type of hyper-personalism in a movement of greater consciousness, always ascending until the completion of Christ's Body in the parousia."[5] In Teilhard's view, Christian life is essential to the progress of evolution. He emphasized that the role of the Christian is to divinize the world in Jesus Christ, to "christify" the world by our actions, by immersing ourselves in the world, plunging our hands, we might say, into the soil of the earth and touching the roots of life. In his *Divine Milieu* he wrote, "There is nothing profane here below for those who know how to see."[6] The world, he claimed, is like a crystal lamp illumined from within by the light of Christ. For those who can see, Christ shines in this diaphanous universe, through the cosmos and in matter.[7] He posited a "mysticism of action" in a universe moved and compenetrated by God.[8] For him, union with God means not withdrawal or separation from the activity of the world but a dedicated, integrated, and subliminated absorption into it.[9] Formerly, he said, the Christian thought that she or he could attain God only by abandoning everything. One now discovers that one cannot be saved except through the universe and as a continuation of the universe. We must make our way to heaven *through* earth.[10]

Teilhard was suspicious that Christianity makes its followers inhuman, presenting them with a series of rote doctrines devoid of life and pointing believers to a starry heaven away from the world. Christians are not conscious of their divine responsibilities, he claimed, but see Christian life as a series of observances and obligations, not the realization of the soul's immense power. This has led to a "static Christianity" that isolates its followers instead of merging them with the mass. A "static Christianity" leads to a routinized or mechanized Christian life whereby the language, symbols, and metaphors of theology and ecclesial life fall into the trap of a misplaced concreteness that resists growth and change. As a result, Christians lose consciousness of their divine responsibilities, and Christianity moves to the margins of sectarianism. If Christians are to remain faithful to the gospel then change is necessary. As Teilhard writes,

> If we are to remain faithful to the gospel, we have to adjust its spiritual code to the new shape of the universe. It has ceased to be the formal garden from which we are temporarily banished by a whim of the creator. It has become the great work in process of completion which we have to save by saving ourselves. We are the elements responsible, at the atomic level, for a cosmogenesis ... [The Christian] is now discovering that s/he cannot be saved except through the universe and as a continuation of the universe.[11]

> Christ must be born again. Christ must be re-incarnated in a world that has become too different from that in which he lived ... I believe that the Messiah whom we await, whom we all without any doubt await, is the universal Christ; that is to say, the Christ of evolution.[12]

The idea that the trajectory of evolution depends on Christian participation in the world impelled Teilhard to envision a vital role for the church. In his view, the church is a phylum of Christian amorization in the universe; it is the source of a new type of humanity—christified humans. Love is the "within" of things (hence, "amorization"), the immanent force unifying all conscious beings, "personalizing" by totalizing. Christian love is the energy of the new evolution because Christ is the center of Christian love. How does a person make contact with Christ, the center of the universe? Teilhard claimed that such contact takes place through the Eucharist. In his *Divine Milieu* he describes how Christ's cosmic activity emanates from the Eucharist to touch each of our material activities: from our sacramental lives in union with Christ to the sacrament of the universe. Through the body of christified persons, Christ reaches humankind and the material universe.[13] Teilhard spoke of the Christian phenomenon as a new Christian phylum that transcends *homo sapiens*. This "phylum of salvation," according to Teilhard, is to spread its inner life and hyperpersonalism "in a movement of greater consciousness, always ascending until the completion of the Body of Christ in the parousia."[14] Christ's transforming activity must move from the church's altar to the altar of the material universe.

He described the church as a "phylum of love," positing a new concept of church that includes the cosmos. Christian love, he claimed, is the energy of the new evolution because love unites and differentiates. Similarly, Bede Griffiths described the purpose of the church as the communicator of love. He writes,

> The organization of the Church, with its doctrine of the Trinity and incarnation and its Eucharistic ritual, has no other purpose than to communicate this love, to create a community of love, to unite all in the eternal Ground of being, which is present in the heart of every person. This is the criterion by which the Church is to be judged, not by the forms of its doctrine or ritual, but by the reality of the love which it manifests.[15]

Teilhard saw a unique role for the cross as the sign of progress toward union in love. The cross does not assume the role of expiation but is the very sign of evolution, as suffering and death yield to greater union. He wrote, "When the cross is projected upon such a universe in which struggle against evil is the sine qua non of existence, it takes on new importance and beauty ... He [Christ crucified] is the symbol and the sign-in-action of progress. The complete and definitive meaning of redemption is no longer only to expiate; it is to surmount and conquer."[16] This optimistic view of the cross with an emphasis on love revealed a certain naïveté in Teilhard, or so his critics thought. Where is the place of evil in his doctrine, they asked? How do we make sense of tragedy in the world if love is the guiding force? One could speculate on evil and tragedy from the comfort of one's desk, but Teilhard pondered evil from the trenches of war. As one caught up in the evil and horror of World War I, Teilhard was privy to evil and tragedy. He experienced the tragic side of humanity firsthand. One of the first essays he wrote during the war in 1914 dealt with his reflections on the problem of evil, a subject continually in his mind. Yet, he understood that beneath the problem of evil (which he hoped to solve) there still lay a "mystery."[17] The "mystery," as Teilhard perceived it, was the forward and upward movement of evolution. What keeps this universe moving in a forward direction despite the resistant forces of evil? If the universe is evolving toward the fullness of Christ, can we speak of resistance to evolution as the Antichrist? On one hand, Teilhard viewed evil as an "evolutionary by-product" resulting from "resistances to the spiritual ascent inherent in matter," which indicates that every stage of evolution involves disorder. He did not intend to dismiss evil or simply tolerate it but to recognize that there is an inner pressure toward evolutionary progress, and, at least in the natural world, tragedy is inherent to this progress. On the level of humanity, however, he emphasized moral freedom. He recognized that the human person must choose between revolt and adoration, between salvation and loss. It is for this reason, I believe, that spirituality plays such a vital role in his doctrine. In a universe marked by freedom, sin is not only the force opposed to freedom but that which makes freedom real.

In my view, it is Teilhard's experience of evil that makes his doctrine credible. His emphasis on evolutionary progress through love is not a romantic dream but the insight of one who saw a power at work within the universe greater than the force of evil, the power of love. The whole evolutionary process could be described as a process of amorization. In his *Phenomenon of Man* he wrote, "If in our love relationship with another we find our truest 'person,' why should it not be true on a world-wide dimension?"[18] With cosmogenesis being transformed into Christogenesis, the very being of the world is now being personalized. "Someone" and no longer something, is in gestation in the universe.[19] According to Teilhard, the Christian vocation now is "no longer merely to ease the suffering, to bind up the wounds or to succor the weak, but through every form of effort and discovery, to urge its powers by love right up to their higher term."[20] We are to harness the energies of the world through the human energy of love, which is the energy of convergence in evolution. Thus, the church is to be the place where Christ amorizes Christians through community and a sacramental life, where Christian love is energized and synthesized for the emerging new creation which is at the heart of the whole evolutionary movement.

If love is at the heart of Christian life in the universe, then "doing Christ" is a reflection of the divine light shining through one's life. According to Merton, the light within brightens by discovering our true identity, which can be found only in God; it is the initial growth of integration and unity. As we come to discover ourselves in God and contemplate the human mystery in the image of the crucified Christ, we come to a reconciliation of opposites within us. Then, and only then, are we prepared to live, like Christ, for the sake of others. Then do we see our neighbor as brother or sister and not as an infringement on our private lives. As we allow the Spirit to search the depths of our own lives, so too we can begin to search the depths of others, the depth of God that is revealed in the human face. Through the grace of the Holy Spirit, we acquire the gifts of mercy, forgiveness, peace, charity, kindness, which help create us as christic persons. As Christ is formed within us, our vision of the world begins to change. We are drawn from that self-clinging which makes authentic relationships impossible to real vision. As Merton writes,

> The familiar phrase "seeing Christ in my brother" is subject to a sadly superficial interpretation. How many Christians there are ... who do not hesitate to assert that this involves a sort of mental sleight-of-hand, by which we deftly do away with our neighbor in all his/her concreteness, his/her individuality, his/her personality ... and replace him/her by a vague and abstract presence of Christ. Are we not able to see that by this pitiful subterfuge we end up by trying to love, not Christ in our brother/sister, but Christ *instead* of our brother/sister? Our faith ... is the needle by which we draw the thread of charity through our neighbor's soul and our own soul and sew ourselves together in one Christ. Our faith is given to us not to see *whether* or *not* our neighbor is Christ, but to recognize Christ in him and to help our love make both him and ourselves more fully Christ.[21]

The inner conviction and experience of God's love impel one to a new level of personhood, as well as a new level of consciousness, which is more relational and attentive to the other because of a more integrated and differentiated self—a christic self. As Paul writes, "The life I live now is not my own; Christ is living in me" (Gal 2:20). To become a christic person is to come to a new level of personhood, a transcendent level, by which we realize that the God of love who permeates our lives, lives in our brothers and sisters as well. It is to come to a new understanding that we humans are not autonomous atoms in the universe; we are not ontologically separate beings. Modern physics has changed our understanding of the world and introduces us to the wholeness of cosmic reality. In the *Tao of Physics* Fritjof Capra writes,

> Quantum theory has abolished the notion of fundamentally separated objects, has introduced the concept of the participator to replace that of the observer, and may even find it necessary to include human consciousness in its description of the world. It has come to see the universe as an inter-connected web of physical and mental relations whose parts are only defined through their connections to the whole.[22]

We belong to an implicate order that integrates the entire universe—a single web of light across the span of time and the levels of existence. The physicist David Bohm writes, "We have to regard the universe as an undivided and unbroken wholeness."[23] To come to this new level of personhood within an interconnected universe is to contemplate Christ, to penetrate the truth of the Christ mystery within ourselves, in other persons, and nonhuman creatures as well. Beatrice Bruteau describes the mystery of Christ in a way that helps us appreciate it in the second axial period. She writes,

> To enter by our transcendent freedom into Christ and to become a New Creation means to enter by faith into the future of every person and into the very heart of creativity itself, into the future of God.
>
> To be "in Christ" is to accept the offer that Jesus makes, to be food for his friends. One must renounce the lordship pattern of organizing social relations. One must forsake being either dominant or submissive. One must undergo this "particular mutation in consciousness." To be "in Christ" is to enter into the revolutionary events of Holy Thursday by ...

letting an old modality of consciousness die and seeing a new one rise to life.

To be "in Christ" is to abandon thinking of oneself only in terms of categories and abstractions by which one may be externally related to others and to coincide with oneself as a transcendent center of energy that lives *in* God and *in* one's fellows—because that is where the Christ lives, in God and in us.

To be "in Christ" is to experience oneself as an initiative of free energy radiating out to give life abundantly to all, for that is the function of the Christ. To be "in Christ" is to be an indispensable member of a living body, which is the Body of Christ.

To be "in Christ" is to be identified with the Living One who is not to be sought among the dead, for the Living One is the One who is Coming to Be.

If I am asked then, "Who do you say I am?" my answer is: "You are the new and ever renewing act of creation. You are all of us, as we are united in You. You are all of us as we live in one another. You are all of us in the whole cosmos as we join in Your exuberant act of creation. You are the Living One who improvises at the frontier of the future; and it has not yet appeared what You shall be.[24]

What Bruteau highlights is that living in Christ is living in relationships—*in* God and *in* one's neighbors—because that is where Christ lives. It is living with a consciousness of connectivity to my neighbor, who is my brother or sister. A cosmic christic world means living in the mystery of Christ as the integrator and unifier of diverse people by first becoming integrated and unified in oneself. Merton wrote: "What we are asked to do at present is not so much to speak of Christ as to let him live in us so that people may find him by feeling how he lives in us."[25] It is living Christ as the mystery of unifying love that renders us co-creators of an evolutionary universe. Love has the power to transform, and the power of transforming love lies within us, for each of us has the power to make present the living God.

To become a co-creator in Christ is to become a vessel of love by which God's selfless love may flow into the world and thus unify the world in love. It is to come to a new level of consciousness and attract a new vision of reality, a "transcultural" and "transsocial" level of consciousness, where we realize that our neighbor is not the exception to God; rather, the neighbor *is* God enfleshed in a fragile form of human weakness. Bonaventure helps us realize that Christian love which integrates and unifies can only come about through the cross, which means self-acceptance, self-love, and self-surrender by which we allow God to be the God of our lives. Sin is living in the exile of unrelatedness. As Bede Griffiths reminds us, "Your own existence is your greatest sin—that is—your separate existence apart from the Self."[26] Sin is the refusal to grow and change, to accept the demands of love and hence participate in Christogenesis. The sinful person refuses the call to community and thus becomes a resistant force to the movement of Christ in evolution.

Because the wages of sin are fragmentation and division, sin can be healed only by a crucified love, the unconditional love of God revealed in the cross of Jesus. Crucified love is love that creates a new future because it is the love that makes space for others to enter in and share life. This love is salvific because it recognizes that healing and wholeness (*salvare*) in the human community require *inner* healing and wholeness. "The root of Christian love," Merton wrote, "is not the will to love, but *the faith that one is loved.*"[27] To love in the spirit of crucified love is to let go of power and control, which separate and isolate, and to give birth to Christ through a self-surrendering love. Beatrice Bruteau writes, "If we really accept that creation is always new, and if we ourselves are active participants in this new creation, then we are always facing the future."[28] The cross signifies that we must let go of separateness and challenge ourselves to a higher love for the sake of union. As the author of the Fourth Gospel reminds us, "Unless a grain of wheat falls to the ground and dies, it remains a single grain but if it dies it produces an abundance of life" (John 12:23-24).

To be a co-creator in Christ by which Christ becomes co-creative in the universe is to recognize the nobility of being human, that is, of being image of God. Every human person bears this image and thus has the capacity to express God incarnate, who is the Christ. The Christian task is to help realize the power of Christ as the unifying center of the universe. As christic co-creators, therefore, we are not to exclude people of other religions but are instead to enter into union with them. Dialogue, as Panikkar and Griffiths point out, is the path to union and hence to the fullness of Christ in creation. Dialogue is not talk; rather, it is conversation insofar as it is a "speaking with" the other, a mutuality of shared ideas. Dialogue is less an exchange of words than an event by which the Word or words that are spoken enter our lives and change us. In this way, interreligious dialogue is a "desert experience." Among the ancients the spoken word was an empowered event, one that connoted some type of praxis. For the desert fathers, the spoken word imparted action and direction to the disciple's life.[29] The power of the word changed, according to Walter Ong, when the word became a printed word and thus a "thing" out there on a flat surface. Ong suggests that such things "are not actions but are in a radical sense dead, though subject to dynamic resurrection."[30] In the same way, interreligious dialogue can be a string of words subject to individual interpretation unless they are received or exchanged in a spiritual openness to new life, that is, unless dialogue takes place in the "cave of the heart." Interreligious dialogue, like Christology itself, must be nurtured by prayer and contemplation. Only through a poverty of spirit and openness of heart can we receive the words of other religions and find in them the overflowing goodness of God and the voice of Christ in other languages and other ideas.

If Christology in the second axial period seeks a rebirth, dialogue may be the best way we can begin to detach ourselves from the identification of Christ with Western culture and open ourselves up to the tremendous mystery of Christ in other religions and nonhuman creation. That is, dialogue may underlie the new "ecology of Christ." Panikkar states that real intrareligious dialogue begins in myself; it is more an exchange of religious experiences than of doctrines. If one does not start from this foundation, no religious dialogue is possible. This too was the spirit of Vatican II, which held "there is no ecumenism worthy of the name without interior conversion" (*Unitatis Redintegratio* 7 [Decree on Ecumenism, 1964]). Yet, John Borelli has recently pointed out that the church is retreating from interreligious dialogue and is less willing to nurture interior conversion as the way to dialogue. This retreat can only contribute to a widening gap between the church and the laity. As Borelli writes,

Whether Catholic bishops care to think ecumenically about the nature of the church, more and more people will do so simply because more families now include members who attend more than one church. Christians are worshiping together in more and more settings, studying Scripture together, praying together in all sorts of groups. Church leaders working outside ecumenical and interreligious contexts are an anachronism in our contemporary world, which John Paul II once described as existing in a climate of increased cultural and religious pluralism.[31]

Although Borelli describes a weakening of interest in interreligious dialogue among the bishops, we must note the efforts made by Pope John Paul II to foster peace among world religions by organizing two major gatherings of religious leaders in Assisi, the first one on October 27, 1986, and the second on January 24, 2002. On the twentieth anniversary of the first gathering, Pope Benedict XVI marked the occasion by affirming that religion must be the herald of peace. He states,

Despite the differences that mark the various religious itineraries, recognition of God's existence, which human beings can only arrive at by starting from the experience of creation [cf. Rom 1:20], must dispose believers to view other human beings as brothers and sisters. It is not legitimate, therefore, for anyone to espouse religious difference as a presupposition or pretext for an aggressive attitude toward other human beings ... Among the features of the 1986 Meeting ... the people of diverse religions who were praying could show through the language of witness that prayer does not divide but unites and is a decisive element for an effective pedagogy of peace, hinged on friendship, reciprocal acceptance and dialogue between people of different cultures and religions. We are in greater need of this dialogue than ever, especially if we look at the new generations. Sentiments of hatred and vengeance have been inculcated in numerous young people in those parts of the world marked by conflicts, in ideological contexts where the seeds of ancient resentment are cultivated and their souls prepared for future violence. These

Panikkar points out that the purpose of dialogue is not to establish some universal religion but to communicate the various ways God is present, incarnate, in creation; to bring Christ alive not by doctrine but by bonds of relationship that help unify the world in God. Panikkar holds that "each religion represents the whole for that particular group and in a certain way 'is' the religion of the other group only in a different topological form."[33] "Although admitting that such a view may sound too optimistic," Gerry Hall states, "it provides insight into the basis upon which homologous correlations can be made. Although religions and cultures are profoundly unique, they may represent transformations of a more primordial experience that make each tradition a dimension of the other. If this is the case, then dialogue may not only uncover hidden meanings within another religious system; it also discovers hidden or repressed meanings within one's own."[34] Interreligious dialogue can proceed only on the basis of a certain trust in the "other qua other," a kind of "cosmic confidence" in the unfolding of reality itself.[35] Hall notes there are certain indispensable prerequisites for dialogue. These include a deep human honesty, intellectual openness, and a willingness to forgo prejudice in the search for truth while maintaining "profound loyalty towards one's own tradition."[36] Panikkar gives the example of the Greek and Christian conceptions of the Logos, which appear conceptually distinct, even contradictory. The former is a semi-divine, created principle of rationality in the universe; the latter, a fully divine, noncreated power in the world. Once these two symbols are thought through together, however, "the former had to offer a certain affinity to the new meaning that would be enhanced once it was assumed."[37] In this way, there is a coalescence of symbols within both traditions. Panikkar's notion of homeomorphic equivalence not only recognizes points of encounter; it equally suggests a process of "mutual fecundation." Religions and cultures continue to intertwine historically and existentially so that self-understandings and symbols are in a constant process of mutual influence and growth.[38]

The evolution of complexified religious consciousness by which the mystery of Christ is in evolution depends on a spirit of openness and receptivity to the other. When doctrine becomes rigid and truth vigorously defended, the whole evolutionary movement is resisted by the force of fear. We become afraid to live in the incomprehensible mystery of God's love. Yet, Christ is the visible living reality of God's love. To love God is to live Christ. From a theological perspective, Christ is the Word of God and, as divine Word, cannot be confined to a religious culture because the fullness of Christ must include every person and all creation. Because Christ is the mystery of the whole creation, the one spoken Word of God incarnate, Christ is more than we can grasp or understand. The risen Christ reveals the victory of God's kenotic love for all people; all are invited into this love through the logic of the cross. Thompson writes, "The transforming Christ summons every religion to a continual conversion and regeneration, and thus the finitude and sinfulness of every religious expression is clearly recognized ... The transcultural Christ is co-present wherever the Divine *Pathos* is authentically mediating itself, whether through some cultural form of religion or through some more 'personal' religious experience."[39] How do we liberate Christology in the second axial period to proclaim Christ—not of the empire but Christ of the cosmos?

We must find a way to unleash the Spirit in our midst, the Spirit that is struggling to create anew. The vernacular theologians of the second axial period point the way not by formal study alone but by study accompanied by prayer and conversion, that is, a development of inwardness. We need a return to prayer, solitude, contemplation, and scripture as the source of spiritual life, and spirituality as the source of theology. We need a penetrating insight into God as creator, as the God of hope and promise, and the future into which we are moving. I am not suggesting that we eliminate academic theology, for it serves a vital apologetic function through critique and rigorous questioning; however, I am suggesting that it be complemented by contemplative theology and a development of inwardness. Contemporary theology has given insufficient attention to many key elements of human life, such as contemplation, silence, and the well-being of the body; yet, these are essential to a renewed cosmic Christology in the third millennium. This is perhaps one of the most challenging tasks facing us today, the rebirth of Christ for the life of the world by the rebirth of Christ within us. The institutional church maintains a level of control that does not get beyond the fixity of dogma. Christians are not encouraged to engage in the mystery of Christ in such a way that a christic consciousness is formed in the ecclesial community. What would the world be like if Christians lived in the liberating Spirit of Christ, evolving toward a higher level of christic consciousness, creating a new world by dying to the old one?

We are confronted today by the survival of Christianity. Will it survive in the face of evolutionary shifts marked by a new global consciousness or is it a diminishing sect in an ever-expanding universe? Teilhard maintained that not only are we to survive but we are to flourish. We are to lead the evolutionary trend in a forward movement into God. He urged Christians to participate in the process of Christogenesis, to risk, get involved, to aim toward union with others, because the entire creation is waiting to give birth to God's promise—the fullness of love (Rom 8:19-20). We are not only to recognize evolution but make it continue in ourselves.[40]

It is precisely the new unfolding life of the universe that impelled Teilhard to see Christianity as a religion of evolution. Christ is the evolver, the centrating energy of the evolutionary movement. But Christ cannot be the energy of evolution unless the incarnation is allowed to continue in us. That is, Christ is in evolution insofar as the life of Christ continues in us—humanity and creation. In this respect, the direction of evolution depends on our participation in Christ. It is no longer enough to pray to Christ—to "be saved." Now salvation means participation in the mystery. Our participation in the Christ mystery is necessary for the fullness of Christ. Christ must come alive in us if this universe is to find its fulfillment in God. What took place in the life of Jesus must take place in our lives as well if creation is to move toward completion and transformation in God. We are to give ourselves to Christ and to his cause and values, which means not losing the world but finding the world in its truest reality and in its deepest relation to God.[41] Zachary Hayes writes, "we are to fill the Christ-form with the elements of our personal life and thus embody something of the Word in ourselves in a distinct and personal way."[42] Our participation in the mystery of Christ, therefore, lies at the basis of a healing world, a world aimed toward the fullness of the reign of God. This "putting on Christ" through the living out of an agapistic ethics means living in creation as gift, relating to creation as family, that is, as brothers and sisters, and treating the world of nature with respect and a healthy concern.[43] Hayes indicates that this evolutionary world can move forward toward its fulfillment only because of our loving actions and not apart from them. He gives a positive emphasis to the role of humans in the mystery of Christ but also indicates that without our participation, creation will not attain its destiny in God.

In order for Christ to live, we must let go of possessing Christ— "do not cling" to me, Jesus said to Mary, but go and announce the good news (John 20:17). The system theorist Erich Jantsch wrote that "to live in an evolutionary spirit means to engage with full ambition and without any reserve in the structure of the present, and yet to let go and flow into a new structure when the right time has come."[44] But it is the letting go that we resist. The contemporary writer Ken Wilber points out that many people are terrified of real transcendence because transcendence entails the "death" of one's isolated and separate sense of self. To strive for wholeness, to live in transcendence, is to let go and die to one's separate self. The dilemma, however, is that we resist the very thing we desire, transcendent wholeness, because we fear the loss of the separate self, the "death" of the isolated ego. As Wilber writes, "Because we want real transcendence above all else, but because we will not accept the necessary death of our separate self-sense, we go about seeking transcendence in ways that actually prevent it and force symbolic substitutes."[45] Could we not say the same on the level of the institutional church? The very ideal we profess—the body of Christ—we resist because of the demands of *kenosis,* suffering, and death, and the risks involved. I would add that we resist divinization. We are created with the capacity for God, but we resist our desire to be like God because we resist conversion—so we create our own gods, which increases our loneliness and separation. As a result, we stay fixed in our tiny places in the universe, defending our private turfs lest they be overturned in order to plant new seeds. New theological ideas or insights challenge our established beliefs, sometimes yielding to embattled arguments; other cultures and religions frighten us into the trenches of fundamentalism; the neighbor becomes an infringement on our private space, and community becomes a burden of rituals and observances. Although we may be unhappy with our estrangement, we remain fixed in position because it is secure (or so we think) and we can control our boundaries. For Christians, estrangement is deadly. Instead of becoming a Christian "phylum of love" (and thus leading the evolutionary process) we live in an abandoned fixity of species—a static

Christian life in a static Christian church. We wind up thwarting the very claim of our existence, the centrality of Christ.

If being the image of God is at the heart of evolution in Christ, then the spiritual life is essential to Christian evolution. A dynamic interior spirit must be at the heart of change. Change is not what happens outside us; rather, change must first take root within us. If we cannot embrace change interiorly, we will not accept change exteriorly unless it is thrust upon us; and then it is not life-giving change but forced endurance. Just as the world of nature has an inner freedom to be itself, so too Christian life, if it is truly an evolutionary life in Christ, must be rooted in freedom. "If anyone is in Christ," St. Paul wrote, "that person is a new creation" (2 Cor 5:14). Conversion is the movement toward authentic freedom, the freedom to be oneself and hence for God (since the source of our lives is in God). It is a way of becoming more authentically human through an interior attitude of "turning" from a selfish self toward a self that is centered on the other. Just as the physical world is comprised of integrally related particles, so too, we are integrally related to one another and to (nonhuman) creation. We live in a web of relationships. As we affect this web by our actions, so too are we affected by it. Conversion is accepting interdependence as the definition of life in the universe.

Christian life in the second axial period must be marked by openness to God, not a transcendent a-historical God but the God of history, the God of evolution. God is the future into which we are moving, not a tribal God of some ancient history. We live in a dynamic and unfolding universe; ours is an open system of life. In light of an evolving universe where change is integral to the emergence of new life, we should welcome change as the very sign of life. To resist change is ultimately to resist Christ; it is to prevent evolution toward unity of life in the universe. To be a Christian is to be "on the way," announcing the good news of the risen Christ through spiritual attitudes of poverty of being, humility, compassion, openness of heart and mind, values that can let Christ live in us and in others as well.

Christology in the second axial period needs a new understanding of Christ, a new way of doing theology, and a renewed sense of Christian life. The vernacular theology of the mystics is the most viable way that Christ can be raised from the dead and become "God for us"—through participation, dialogue, and engagement with the world. Teilhard's spiritual vision, centered on and rooted in Christ, emphasizes "global responsibility, action and choice in shaping the future of humanity on our planet. He affirms that life is a task to be done, a work to be achieved, and celebrates life as a most precious and wonderful gift to be loved and experienced as a sign of the Spirit who sustains us all."[46] This is the insight of a Christian deeply in love with God, one who nurtured the mystery of Christ deep within his being and gave birth to Christ in his life, despite opposition by his religious order, by other theologians, and censorship by the institutional church.[47] Teilhard, Panikkar, Merton, and Griffiths are icons of Christ and models of holiness in the second axial period, for they urge us to let go and die into love— to be what we say we are, Christian. They show us that fear is driven out by perfect love. They were not afraid to live as Christians and neither should we be. Like them, we too are to seek the hidden God in our world by seeking the hidden God in our lives—living Christ by doing Christ. Perhaps another great mystic of our time, Karl Rahner, said it best: "The Christian of the future will either be a mystic, one who has experienced something or she will cease to be anything at all."[48]

CHRIST IN EVOLUTION: TECHNOLOGY AND EXTRATERRESTRIAL LIFE

The renewal of Christ in the second axial period is a renewal of life for humanity and for the earth itself. This renewal is a present reality and a view toward the future where the fullness of Christ is envisioned as a world of many peoples, religions, cultures, and languages united with God and with one another. Eschatology is a way of speaking of the future from the basis of religious experience. As a Christian mode of discourse, it is a projection of fulfillment from the present experience of the human situation insofar as the present is already conditioned by the mystery of the Christ-event. The resurrection of Jesus is the beginning within history of a process whose fulfillment lies beyond history, in which the destiny of humanity and the destiny of the universe together find their fulfillment in a liberation from decay and futility. The Christian claim is that Jesus Christ really lives with God. Since the resurrection of Jesus anticipates the destiny of humankind, Christian theology affirms that the whole creation has a future in God.

Teilhard de Chardin was convinced that the total material universe is in movement toward a greater unified convergence in consciousness, a hyper-personalized organism. He conceived of the universe as a vast transhuman body in the process of formation, held together by the Omega point, "a distinct Centre radiating at the core of a system of centres."[1] Because of Christ, Teilhard indicated, we live in "an irreversible personalizing universe."[2] In the light of this personal universe, Teilhard spoke of the organic nature of Christ as the total Christ whose activity consists in "recapitulation," or bringing the universe to its ultimate center through the transforming energies of the resurrection. Christ is the physical center of an expanding universe. By "physical" Teilhard meant ontological reality. Christ is the real personal center of the universe.

156

Rahner too expressed this idea when he wrote that Jesus, in his death, became a permanent ontological determination of the whole of cosmic reality. Denis Edwards writes, "The salvific reality of Christ, which is consummated in death, is built into the unity of the universe, so that the world is a radically different place from what it would have been if Christ had not died."[3] The humanity of Jesus is the permanent openness of finite created being to the living God;[4] the resurrection is truly the beginning of transformation of the world in God. In Teilhard's view, the world is not blindly hurtling itself into aimless expansion, a mass of heterogeneity, but is moved by Christ to Christ that God may be all in all. Christ is the goal toward which the whole cosmos is moving and in whom the cosmos will find its completion. In his grace-filled human reality, Christ becomes a power shaping the whole cosmos. While this power is real, Christ also symbolizes the potential within humanity and the cosmos for union with God. As we discussed in the last chapter, the future of the material universe is intimately linked to the fulfillment of human beings in whom the world has come to consciousness. What we do matters to the "matter" of the universe, because by our choices we influence the life of the universe.[5]

In this respect, the consummation of the universe, the parousia or second coming of Christ, will ultimately be determined by the choices of the human community. Panikkar writes that the parousia of Christ is not separate from the Eucharistic and risen Christ; it is not another incarnation or a second Christ appearing somewhere. We have been warned by Jesus in the Gospels not to believe in any appearance of the coming Messiah here or there. Rather, we are co-workers with God and stewards of creation. The second coming of Christ is the emergence of Christ in us, the human community, when we become reflective not only individually but collectively and live in the spirit of crucified love. Jesus came once; now the new Adam, the new earth, must be fully formed in us if this universe is to find its completion in God. Bonaventure also said that the final age of peace is yet to come. In this final age, Christ will appear and suffer in his mystical body, which indicates that crucified love will mark the final age.[6] As George Maloney points out, the progress of the universe cannot be in competition with God. "The world cannot have two goals, two ends, two summits, two centers. Christ alone is the center and the goal of our universe."[7] Teilhard indicated that "the total Christ is only attained and consummated at the end of universal evolution."[8] The one Christ who emerges out of an evolutionary universe and who is the center and goal of this universe will come to completion only at the end of evolution, that is, the Christ of the physical universe, the Christ of all humanity, the Christ of all religions. In this respect, Christ is not a static figure, like a goal post with a gravitational lure, toward which the universe is moving. Rather, Christ is in evolution because we, human and nonhuman creation, are in evolution. If our actions and choices influence the building up of Christ in the universe then we must take seriously the impact technology and science are causing on the shape of life in the universe.

The new science, as we indicated in chapter 8, portrays reality not as individual, mechanistic parts but as an interrelated whole. Fritjof Capra claims that a wholistic worldview recognizes the intrinsic value of all living beings and views humans as one particular strand in the web of life.[9] If the interrelated whole, including the physical cosmos, characterizes life in the universe, then what does this mean for the future of life? Some scientists believe that the universe will continue to expand forever. Others claim that it will eventually collapse, although this will not take place until another ten thousand million years pass. Still others say that the universe is oscillating between cycles of contraction and expansion. The "big crunch" will stop when the contracting universe reaches a certain density and then will "bounce" back into a cycle of expansion.[10] Whatever the future of the universe will be, the world *as we know it* will eventually come to an end. This has impelled the world of technology and the engineers of artificial intelligence to seek means of prolonging life in the universe. According to Freeman Dyson, "Life resides in organization rather than in substance."[11] If organization can be achieved through any material substrate, he claims, then computation by appropriately organized matter can beat the second law of thermodynamics, which states that we are wearing down in the universe. Scientists tell us that biological evolution may be winding down but technological evolution has started to accelerate.

Technology can be defined as the organization of knowledge for the achievement of practical purposes. We may also describe it as the development of mechanical devices by the human community in its efforts to control or exploit the forces of nature.

Throughout history, humans have been inventive in various ways, enhancing human life through means of technology. The universe has been an instrument for the gradual evolution of humans to full self-consciousness, self-development, and self-expression of free conscious spirit in and through matter. The development of technology expresses the human's self-development and self-expression through matter; it is integral to being the image of God and thus integral to authentic self-realization.[12]

The notion of a dynamic image of God leads to the deeper significance of cultivating science and technology. The human person has become transformed by his/her own initiative and artistic inventiveness which expresses in a new way both the divine image of its creator and the human image of its re-creator. The notion of the human as a dynamic image of God, with a vocation to develop this image by an evolving dialogue with the material cosmos, sets technology in a wider framework that provides strong religious, moral, and humanistic controls on its exploitation. David Noble claims that technology should have as its aim the unfolding life of the spirit, because technology and religion are similar. Teilhard held that "the technical world of man's daily activities is the 'stuff' out of which the community in Christ is formed to give adoration through the whole universe to God."[13] The incarnation gives Christians the perfect archetype and model of openness of matter to spirit and matter's ability to serve as the medium of the spirit's self-expression and creative power. Modern technology and religion have evolved together, and, as a result, the technological enterprise has been and is suffused with religious belief.[14] Both technology and religion are working toward the new creation because they both have an enduring other-worldly quest for transcendence and salvation. But has technology come to embody our chief values, the things we most want in life?

The roots of modern technological advancement are in Western consciousness and can be traced to the time when the useful arts first became implicated in the Christian project of redemption. The worldly means of survival were turned toward the other-worldly end of salvation, and over time the most humble of human activities became invested with spiritual significance and a transcendent meaning—the recovery of humankind's lost divinity. Noble states that when people wonder why new technologies rarely meet human and social needs they assume it is because of greed and lust for power. He claims that, on a deeper level, these technologies have not met basic human needs because that is not their primary purpose. They have been aimed rather at the goal of transcending mortal concerns altogether. The engineers of technology today are "offering salvation by technological fix," while making the world over to their vision of perfection. He concludes that the technological pursuit of salvation has become a threat to our survival.[15] This threat can be perceived when we attend to futurists such as Ray Kurzweil, who calls this present era "the singularity" because it is a merger of human intelligence and machine intelligence in which machine intelligence appears to biological humanity as its transcendent servant.[16] The development of intelligent machines that can do all things better than humans is the core of Kurzweil's prediction. In his view, we identify more with our brains than with our bodies.[17] In an interview with *Edge* magazine, Kurzweil stated,

> We are entering a new era. I call it "the Singularity." It's a merger between human intelligence and machine intelligence and is going to create something bigger than itself. It's the cutting edge of evolution on our planet. One can make a strong case that it's actually the cutting edge of the evolution of intelligence in general, because there's no indication that it's occurred anywhere else. To me that is what human civilization is all about. It is part of our destiny and part of the destiny of evolution to continue to progress ever faster, and to grow the power of intelligence exponentially. To contemplate stopping that—to think human beings are fine the way they are—is a misplaced fond remembrance of what human beings used to be. What human beings are is a species that has undergone a cultural and technological evolution, and it's the nature of evolution that it accelerates, and that its powers grow exponentially, and that's what we're talking about. The next stage of this will be to amplify our own intellectual powers with the results of our technology.[18]

Kurzweil is not alone in his convictions. Some scientists see that the ultimate goal of artificial life is to create "life" in some other medium, ideally a virtual medium in which the essence of life has been abstracted from the details of its implementation in any particular model. One might say that the ultimate goal of artificial life is to achieve an "artificial reincarnation" of the human species.[19] Hans Moravec claims that the advent of intelligent machines (*machina sapiens*) will provide humanity with "personal immortality by mind transplant." Moravec suggests that the mind will be able to be downloaded into a machine through the eventual replacement of brain cells by electronic circuits and identical input-output functions. In this way it becomes possible to transfer a mind from one support to another and hence to bring about the survival of the soul after death in a new, more durable medium.[20]

Noreen Herzfeld claims that the drive to develop artificial intelligence (A.I.) is related to the human *imago Dei* and the desire to create an "other in our own image."[21] She asks, "Is this image that humans share with God related to the image we wish to share with our own creation in A.I.?"[22] In the movie *A.I.*, for example, the robot boy David desires to be human, but discovers that being human means to be mortal. When he finally fulfills his dream to be a real human boy, he dies. Philip Hefner notes that a number of A.I. movies place an emphasis on death and suggests that a good deal of our technology seems to be a denial of death and an attempt to escape it.[23] His insight corresponds to what we might call the A.I. pursuit of "cyber-immortality." Is A.I. pushing us beyond being "human" toward *techno sapiens*? And, if so, how does this affect our relationship with God? Can *techno sapiens* image God or bear a relationship to God?[24] It is interesting that the movie *A.I.* highlights not logic but love as the mark of human life. Robot boy David is programmed to be imprinted to love forever the person who imprints him. The crux of the story is David's relentless pursuit to become a human so that his mother will love him in return. When at the end of the movie he lies next to his mother and hears her say those magic words, "I love you, David," he knows that he has at last become a real boy. Authentic human love cannot be programmed because it is free. It must be freely given and received, and it is in the relationship of reciprocity that love becomes transformative. The receptivity of love for David became the fulfillment of his life—to be human. Death signified the fulfillment of David's hopes and desires—to be human so as to be fulfilled in love. Is this not the meaning of Jesus Christ—to love unto death for the sake of life?[25]

Movies such as *A.I.* remind us that authentic human selfhood is vulnerable to technological subversion and the human obsession with experience. When the self relates impersonally to the world, when things in the world serve as mere objects, the "I" gives expression to an "I-it" relationship. Here the self is concerned with pursuing goals and objectives. The primary concern is to control and manipulate, to treat the world, things, and people as a means of satisfying personal needs and desires. One who lives only with an "it" is not human. To realize one's selfhood, one must enter into a relationship with a "Thou," another person. As Martin Buber wrote, an "I-Thou" relationship requires mutuality, responsiveness, and personal involvement. No one can force another or manipulate another into an "I-Thou" encounter; it cannot be staged. It can only happen in a free relationship in which each person in relationship shares fully and mutually. God is the one "Thou" that does not become an "It." God is the one who gives final meaning to our encounters and reminds us of the "other" as person. Thus, authentic personal relations are beyond the technical sphere.

Since Buber's famous *I and Thou* was written almost fifty years ago, the technological world has infiltrated human life to the point of daily dependency on machines. The influx of technology into daily life is transforming our patterns of play, work, love, birth, sickness, and death. We have become "borged." The *cyborg* is a term of and for our times that maps contemporary bodily and social reality as a hybrid of biology and machine. A cyborg is a person whose physiological functioning is aided by or dependent on a mechanical or electronic device. In a more general sense, cyborg refers "to creatures that are both organic and technological."[26] An increasing number of people are becoming "cyborged" in a technical sense, including people with electronic pacemakers, artificial joints, drug implant systems, implanted corneal lenses, and artificial skin. Still more participate in occupations that make them into metaphoric cyborgs, including the computer keyboarder joined in a cybernetic circuit with the screen, the neurosurgeon guided by fiber optic microscopy during an operation, and the teen game player in the local video-game arcade. The integration of technology in daily human life around the globe has rendered us technologically dependent human beings for whom technology organizes contemporary life. The cyborg is becoming the interpretive symbol for the human self because the cybernetic loop generates a new kind of subjectivity. Naomi Goldenberg is concerned that some Westerners are responding to machines not as tools to use but as role models to emulate. This is leading to an increase in the isolation and loneliness of modern life. She writes, "We are in a process of making one another disappear by living more of our lives apart from other humans in the company of machines."[27]

The prevalence of cyborgs raises questions for Christian life, especially if we claim that relationships of mercy, forgiveness, and compassionate love are the marks of this life. What impels us to act as Christians in a cyborg world if the machine we relate to can satisfy our needs and wants without reciprocal demands? How is Teilhard's vision of a christic universe realized in a species becoming increasingly "techno-sapien"? The challenge of technology to theology is for theology to affirm its own counter-project of life-in-community. The major task for the believing community is to *be* community, to discover and uncover its own reality as the people of God. Technology reinforces the individualistic tendencies of modern society, and it must be given a focus, a direction, because the survival of humanity in a natural environment looms as the issue of our time. Yet, it is not enough simply to come to terms with the integration of technology in daily life. Rather, we must begin to see technology as integral to the whole evolutionary process because it has driven us to a whole new level of culture and consciousness. Hefner states, "Humans are not what they used to be, so anthropocentrism is not the same either. The religion of cyborgs and technosapiens, therefore, is also a religion of nature ... Technology is now a phase of evolution, and it is now creation, a vessel for the image of God."[28] We may speak of Christ, therefore, as the symbol not merely of human life but of cyborg life. If Christ is indeed the goal of this universe, then we must begin to consider this goal in view of the emergence of christic *techno sapiens.*

In a series of lectures on technology and human becoming, Philip Hefner indicated that if we are to speak of religion, it must be a religion that can encompass cyborgs and technosapiens. It cannot be merely a religious way of dealing with technology, as if it were external to who we are; rather, technology has become part and parcel of who we are.[29] When we participate in this drive for new possibilities through technology, we participate also in God. This is the dimension of holiness in technology. The difference between a nontheological and a theological interpretation of technology, Hefner claims, is that the one says the transcending drive is epiphenomenal, a surface phenomenon, while theology says it is rooted in the very nature of things. The epiphenomenalist says that transcendence is evolution's way of promoting fitness. The theologian asserts that evolution has itself been designed to enable a self-transcending system of reality. In Hefner's view, the union of technology and humanity—the emergence of *techno sapiens*—is integral to the transcendent evolutionary trend. Hefner asks, "When we are immersed in the drive for transcendence, are we thereby sharing in the ultimate depths of reality that we can call God? Or are we participating in a process of selection and advance that is finally without meaning?"[30] If evolution is a movement toward greater complexity and union, then the marriage of human creativity and its offspring is integral to evolution. We may view cyborgs in terms of emergence in creation, something totally new that cannot be reduced to its parts, but then cyborgs become integral to the building up of Christ in the universe.

In Teilhard's view, to immerse ourselves in creativity, and thus the drive for transcendence, is to immerse ourselves in Christ, *if faith in the world is faith in God.* That is, if we believe that incarnation means God's involvement in the world, then we must also believe in God's fidelity to the world, which means letting the world become itself through its own creative processes, which now include technology and artificial intelligence. Hefner suggests that we must discover whether there is a deeper meaning to the struggle for transcendence, and we must construct and create that meaning. Technology may be a new expression of Augustine's "restless heart," he states, so that the pursuit of technology is a pursuit of transcendence (toward God) and thus may be a means of grace.[31] Technology is either pointless in the long run (another machine) or it is an expression of the fundamental self-transcending reality of God. Hefner concludes that technology is itself a medium of divine action because technology is about the freedom of imagination that constitutes our self-transcendence. In this respect, technology is integral to evolution. It is the result of human creativity and in turn has influenced the shape of human life to the point of becoming something totally new in the universe.

The human person today is not the same as the human person in the days of Jesus of Nazareth. We are new technological beings, cyborgs, on the way to God. If Jesus the Christ is the symbol of human life then we must admit that the meaning of Christ includes cyborgs; we can now speak of "techno-incarnation" and "techno-Christian" life. The meaning of Christ in evolution includes our ability to transcend ourselves through the creativity of technology so that the shape of Christ is now influenced by technology as well. That is, the human person/cyborg is now techno-transcendent and the future fullness of Christ will be created by technosapiens, if we do not fall prey to immanentism or a metaphysics of materialism. The power of technology is such that it can abet the movement toward the fullness of Christ or it can thwart it. Unless we truly relinquish our freedom to a machine and substitute function for relationship we humans, even technohumans, will be completed in love because that is the meaning of Christ.

While technology is influencing the evolution of Christ in the universe insofar as it is changing the contours of daily human life, there is another dimension of technology that is altering our view of Christ in the universe, that is, the possibility of intelligent life on other planets. The new science of space technology and astrobiology has rendered the discovery of extraterrestrial life more possible. Conversations in religion and science today are beginning to focus on extraterrestrial (ET) life, because such life truly has religious implications, especially for Christians who believe in the incarnation. It is interesting that in 1918 Teilhard wondered whether we ought not to admit a "polymorphic" manifestation of the cosmic Christ in various worlds. Perhaps the human Christ, he said, is simply a "face" of the cosmic Christ.[32] A cosmocentric view of reality commits us to hold that, besides humankind, Christ must redeem any other spiritual beings who may exist in the universe— and the probability that there are many is overwhelming. Teilhard wrote, "Our situation calls for a new Nicaea, to work out a subdistinction in the human nature of Christ, between a terrestrial nature and a cosmic nature."[33] The late Carl Sagan described an attitude of "earth chauvinism," which makes the earth the standard for everywhere.[34] Andrew Burgess identified this chauvinism as a "cosmic hubris"; it "believes that the creator of the universe was incarnate on our own bit of interstellar debris and maintains a special relationship to the human species."[35] As Burgess and others have pointed out, the Christian story is based on a pre-Copernican cosmology. He writes, "As long as someone is thinking in terms of a geocentric universe and an earth-deity, the story has a certain plausibility ... As soon as astronomy changes theories, however, the whole Christian story loses the only setting within which it would make sense. With the solar system no longer the center of anything, imagining that what happens here forms the center of a universal drama becomes simply silly."[36] What Burgess identified is a relationship known for centuries among patristic and medieval writers, namely, theology is integrally related to cosmology. Once cosmology changes, so too does theology, whether implicitly or explicitly.

According to leading scientists, there is mounting evidence that Earth might not be the only oasis in the cosmos or even in our solar system. Researchers have long known that tiny fragments of one planet, such as meteors, can crash onto another planet's surface. This gives life the potential to jump from place to place in our solar system. Recent research suggests that the ingredients necessary for life as we know it to evolve from scratch are present in many places besides Earth. The ingredients include liquid solvents, such as water; a variety of elements, such as carbon, nitrogen, and phosphorus, for constructing organic molecules; and compounds that store energy. According to Bruce M. Jakosky, director of the Center for Astrobiology at the University of Colorado in Boulder, the planet Mars and Jupiter's moon Europa offer tantalizing combinations of these ingredients right in our celestial neighborhood.[37] Lewis Ford writes, "Even if only one planet out of every 150,000 contained life, there would be one million life-worlds in our galaxy, some of which we can reasonably assume contain intelligent life, for whom, we presume, God would also be concerned."[38]

Almost fifty years ago, J. Edgar Burns viewed the widening horizon of the universe through space exploration as the emergence of a new "space-faith" or what he called "cosmolatry." He posited, somewhat prophetically, that "what seems to have received scant attention is the appearance of a phenomenon which has all the earmarks of a new religion."[39] By "new religion" it is not entirely clear whether Burns meant an annihilation of the past or a breakthrough in religious consciousness that renders our "truth claims" obsolete. It is in light of this insight that "exoChristology," a term Burgess used to discuss Christological issues "raised by discoveries in outer space,"[40] takes on new import for Christian faith. For if Jesus Christ is savior, what does this mean for intelligent beings inhabiting a far corner of the cosmos? Or, as Ted Peters asks, if ET life is discovered, would missionaries be called for?[41]

The question of exoChristology is at first disarming because it seems to debunk the salvific work of Jesus Christ. As we discussed in chapter 3, Western Christology following Anselm of Canterbury maintains that human sin is the principal reason for the incarnation. The sin of Adam is repaired by Christ, because what is not healed cannot be redeemed. But as we have argued from the beginning, the reason for Christ is more than sin, it is love. Christ is first in God's intention to love. Hence, Christ is the head of *all* creation, wherever there is life in the universe, as we read in the letter to the Colossians (1:16-17): "All things were created through him and for him. Before anything was created, he existed ... and he holds all things in unity." Christ therefore belongs to the very structure of the universe, and we must begin to consider Christ not just the Omega point of terrestrial life but of the entire universe.[42] If evolution is an unfolding in the universe toward complexity and union, the consideration of Christ and ET is another way of speaking about Christ in evolution.

Bonaventure's integral relationship between creation and incarnation signifies that the incarnation is not an isolated event but is integral to the possibility of creation itself; one is inconceivable without the other. Because of this integral relationship between creation and incarnation, "a world without Christ is an incomplete world," that is, the created order is structured Christologically.[43] Since any created order or all possible worlds are centered in the one Word of God, it could be said that all created orders are structured Christologically. We might term this idea the "christophic principle" because the universe is not simply fit or, even better, fine-tuned, for human life (anthropic principle) but it is patterned on the divine Word of God and thus oriented toward Christ or the perfect union between God and created reality.[44] The christophic principle means that Christ is the reason for every created order, however that created order is conceived, because Christ is first in God's intention to love and thus to create. Because the universe is structured Christologically, every created order will bear a potency for union with the divine. Bonaventure reflected on the perfection of creation in light of the incarnation and said that (in this creation) matter tends toward spirit. The union of matter and spirit in union with the divine is the perfection of creation. Jesus is the fullest realization of the noble potency of creation who brings the created order to its completion. Bonaventure wrote, "The perfection of the entire order is realized for in that one being the unity of all reality is brought to consummation."[45] On an extraterrestrial level, we can speculate that the same divine Word is the pattern of all that exists, imparting to that creation a spiritual potency for union with the divine. Just as Jesus is the fullest realization of divinized humanity in terrestrial creation, so too, the fullest union of natures on an extraterrestrial level will occur (or occurs) in a personal union as well.

Zachary Hayes indicates that the congruent relationship between incarnation and (earthly) creation is the fundamental relationship that renders the cosmos more than material reality; rather, the material world is spiritually potent because it is Christologically structured. The meaning of creation centered in Christ is the mystery of the Word incarnate, grounded in the mystery of divine, self-communicative love. In the context of medieval theology, this is a rejection of the idea that God first created a world that had no relation to the figure of Christ, and that only after the fall of humanity did a "second decree" of God direct itself to the figure of a savior in the form of Christ.[46] Bonaventure and Scotus maintained that a world without Christ is an incomplete world. The whole creation is made for Christ. In short, the primary reason for the incarnation is related not to the "forgiveness of sin" but to the completion of creation in its relationship to God. The incarnation is salvific because all of creation has a spiritual capacity for God which stands open to completion in the human person and ultimately in Jesus Christ.[47]

The Franciscan view of incarnation as a mystery of orderly love uncouples the primary reason for the incarnation from sin and proclaims it as a work of love. An understanding of the incarnation as an act of love rather than a condition of sin may be more fitting to an evolutionary universe where the understanding of human original sin is under revision.[48] Although Bonaventure and Scotus defined God's creative activity in view of *this* world order (since it exists), the theological significance of God's creative action through the Word means that any created reality, wherever it exists, would possess an inner constitution in relation to the divine Word of God. Creation (however conceived) and incarnation belong together. Since Christ is first in God's intention to love, the existence of any world order as an expression of that love will be Christologically structured. Hence, because the incarnation completes that which God creates, extraterrestrials, if they exist, will be open to an incarnation and, in the broadest sense, "saved," insofar as an incarnation of the divine Word will complete extraterrestrial creation in whatever way that creation can fully accept the Word into it.

While the term "incarnation" might not be appropriate to another world order, since it means literally taking on flesh, what we are really talking about is embodiment of the divine Word in created reality. Incarnation as the full self-communication of God's Word in created reality takes place where there is the ability to grasp the divine Word in created reality as the Word of God's love. To say that Jesus Christ reveals God is to say that, in Jesus, God can be known and loved. Incarnation on an extraterrestrial level could conceivably take place, as long as there is some type of intelligence within the extraterrestrial species to grasp the Word of God through knowledge of the divine embodied Word. Since incarnation is integral to creation, intelligibility of the divine Word cannot be divorced from the material expression of that Word. Knowledge of God must ultimately yield to love of God, because God is love. Thus, knowledge of God that leads to a deepening of love must lead to the expression of that love in some form of embodied life.[49] On the terrestrial level, Jesus Christ assumes a bodily nature by which all of creation (that is, material reality) is assumed into relationship with God. Similarly, on an extraterrestrial level, incarnation must assume a form that includes the material reality of that creation, in whatever way that creation is constituted.

The integral relationship between incarnation and creation in view of exoChristology connotes the idea of "many incarnations but one Christ." This idea finds support in the thought of Karl Rahner, who wrote, "In view of the immutability of God in himself and the identity of the Logos with God, it cannot be proved that a multiple incarnation in different histories of salvation is absolutely unthinkable."[50] The depth of Rahner's insight is given breadth and width by Paul Tillich, who argued that the understanding of Christ for the meaning of the universe relates to the understanding of the meaning of the symbol "Christ." By speaking of Christ as a symbol, Tillich challenges us to consider the cosmological significance of Christ. He writes,

> A question arises which has been carefully avoided by many traditional theologians, even though it is consciously or unconsciously alive for most contemporary people. It is the problem of how to understand the meaning of the symbol "Christ" in the light of the immensity of the universe, the heliocentric system of planets, the infinitely small part of the universe which man and his history constitute, and the possibility of other worlds in which divine self-manifestations may appear and be received.
>
> ... Our basic answer leaves the universe open for possible divine manifestations in other areas or periods of being. Such possibilities cannot be denied. But they cannot be proved or disproved. Incarnation is unique for the special group in which it happens, but it is not unique in the sense that other singular incarnations for other unique worlds are excluded ... Man cannot claim to occupy the only possible place for incarnation.[51]

Rahner and Tillich indicate that Christ is not simply the individual existent Jesus of Nazareth but the permanent openness of our humanity to God and hence God's life in us. Tillich's search to "understand the meaning of the symbol 'Christ' in the light of the immensity of the universe" does not diminish the reality of Jesus Christ but restores the meaning of Christ to the person of Jesus who brings terrestrial creation to its fulfillment in God. Jesus is the Christ, and Christ symbolizes the real union of God and creation for terrestrial life; however, as Raimon Panikkar notes, the meaning of Christ is beyond the man Jesus of Nazareth. As we discussed previously, Christ's reality is not exhausted with Jesus' historicity;[52] rather, "Christ is that central symbol that incorporates the whole of reality."[53] John McKenzie states that "Christ is a principle of grace and virtue and these things are fulfilled in Christ: the love of God, the grace of God, spiritual enrichment, the righteousness of God, freedom, strength, faith and love."[54] McKenzie highlights the idea that Christ is first and foremost the life of God by which grace and virtue is possible. Christ is the life not only of human life or life in terrestrial creation but the life of the universe and all universes where there is intelligent life. Christ symbolizes the perfect union of God and creation, which is expressed in a creaturely personal union; thus, the symbol of Christ mediates not only what is symbolized, the life of God, but it symbolizes the life of the creature in God.

In view of the meaning of Christ, we can say that Christ is the symbol of what is intended for created reality, that is, the divinization of creation which, on the level of human experience, reaches its culmination in the person Jesus of Nazareth. Jesus is not the great exception to terrestrial life but its fulfillment. Karl Rahner claimed that, in Jesus, we experience the fact that the mystery of the human person, which is not for us to control, and "which is bound up with the absurdity of guilt and death is, nevertheless, hidden in the love of God."[55] Jesus the man is truly God and the true divinization of our humanity; in him we attain full humanization.[56] However, Christ is more than Jesus. Christ, the Word incarnate, is the one in whom all created reality is transformed in the personal love of God. To speak of Christ on the level of extraterrestrial life, therefore, is not to restrict the discussion to Jesus alone but to see Christ as the icon of created reality. Christ is the divinization of created reality in whatever way the divine Word can fully enter into that reality.

By viewing the incarnation as the primal expression of the love of God, I would suggest that every created life-bearing order is Christologically structured so that, following Rahner's lead, there may be multiple incarnations but one Christ. The reality of Christ is the personal union of God and creation and, as symbol, mediates the divinization of every created order in its relation to God. The cosmological significance of Christ means that Christ is the Alpha and the Omega, the beginning of every created order as Word of God and the completion of that order as incarnation of the Word. In short, Christ, through a self-giving act of love, completes every possible world order by entering into that order through an incarnation or Word-embodiment.[57] The Word character of all possible worlds renders each of them a spiritual potency for union with the divine. The embodiment of the Word in created reality means the ability to express the Word in an intelligible manner and thus to realize the spiritual potential of that created order. The incarnate and risen Word, the Christ, transforms every created life-bearing order in its relation to God. Thus, the meaning of Christ extends to extraterrestrial life because every created intelligible order is disposed to an incarnation by which the Word can be embodied in that creation and transform it from within. However we conceive of the incarnate Word as the integrating and personalizing center of *all* life in the universe (or universes), there remains only one Spirit and one Christ to the glory of God the Father.

CONCLUSION

In a short essay entitled "The Revolution of Jesus," the Camaldolese monk Bruno Barnhart raises the question, "What is distinctive about Christianity?" Or, "What is really new in the Christ-event that was not already present beforehand, either in the biblical tradition or in other religious traditions of the world?"[1] In a sense, we have asked the same questions here but with an emphasis on Christian life. What is the meaning of Christ for Christian life? Does professing belief in Christ make a difference? These questions have been raised not to affirm that Christ is the basis of Christian life but because the meaning of Christ seems to have dropped out of Christian life today and is becoming obscure in the face of religious pluralism.

The belief that Christ signifies a new relationship to God in the world is a fundamental one. For too long, the Christian confession of Christ has rendered Christ a static figure who looms in judgment over the world. Christ has become less a way of life than a law of life. This static figure of Christ stands as a resistant force behind the idea of Christ in evolution. And herein is the problem as I have presented it in this book. The static world of the Greek/medieval cosmos was the world in which the traditional understanding of Jesus Christ was conceived. In this world, marked by order, hierarchy, and structure, Jesus Christ was portrayed as a single, individual existent, born in absolute space and time. A static, ordered, and hierarchical world gave birth to a static, ordered, and hierarchical understanding of Jesus Christ as God and human. While the union of divinity and humanity is not the question here, the question is precisely the meaning of Jesus as the Christ. Through a labyrinth of Greek cosmology, terminology, ecumenical councils, and political battles, we have locked the mystery of God into a single, individual human person, Jesus of Nazareth, so that Jesus Christ has become a single, individual superhero and we are mere spectators to the divine drama. But Jesus is the *Christ*, the Messiah, the anointed one, the one raised from the dead and transfigured in glory. It is difficult to really grasp the significance of Jesus Christ for Christian life if our confession of faith is governed by a medieval cosmology and a fixed place for fallen humanity. The evolutionary worldview, however, has opened up for us a whole new meaning of humanity and, within humanity, the emergence of Jesus Christ. John Haught has spoken of evolution as Darwin's gift to theology, and I think we can say here that evolution is, in particular, Darwin's gift to Christology. For the whole concept of evolution has liberated Christ from the limits of the man Jesus and enabled us to locate Christ at the heart of creation: the primacy of God's love, the exemplar of creation, the centrating principle of evolution, and the Omega point of an evolutionary universe. Rather than fixing our attention on a lonely, static figure of Jesus Christ, we can now locate Christ at the heart of the whole evolutionary process: from cosmic evolution to biological evolution to evolution of human consciousness and culture. Within the evolution of human consciousness, Jesus emerges as the Christ, the fullness of God's self-communication in history and the absolute expression of that self-communication in love. Theologians such as Rahner and Panikkar have helped us understand that Christ is the dynamic life of the world. Christ symbolizes the fullness of life in God. Jesus *is* the Christ because in Jesus the fullness of God's self-communication has been given; God's promise for the world is realized. Because Jesus of Nazareth is the truly human one in whom the fullness of divine communication is expressed, Jesus is the symbol of every person in relation to God. Like Jesus, every person and creature is intended to rise from the dead and share in the glory of God. Jesus the Christ is the realization of what we hope for in the universe: union and transformation in God.

However, Christ is more than Jesus. Christ is the Word incarnate, the one through whom all things are created and in whom all things will find their fulfillment in God. Every person and the whole universe finds meaning in Christ, who bears a distinct and unique relationship with every being; everything is related to Christ. Thus, Christ is the union of humanity and creation with God, the true integrating center of the universe. Because the meaning of Christ is related to the whole—human and nonhuman creation—it is reasonable to speak of the "cosmic nature" of Christ, insofar as one understands "nature" to be defined not by substance (a "thing" in itself) but by relatedness (the Word through whom all things are made). Teilhard wrote, "His birth and gradual consummation constitutes *physically* the only definitive reality in which the evolution of the world is expressed; there we have the only God whom we can worship sincerely."[2] The Christ-event is an organic event; it is living, dynamic, and related to the physical and spiritual life unfolding in the universe. The universe holds together in Christ, as St. Paul wrote, "In Him all things hold together" (Col 1:17). Hence, the meaning of Christ cannot be considered apart from the material universe and its spiritual desire for God.

If Christ is in evolution toward greater complexity so that the *mystery* of Christ is no longer confined to an individual (Jesus) or a local culture or tribe (Christians), then such knowledge of Christ can come about only through a relationship with Christ. In Merton's view such a relationship begins with a life of prayer and an awakening to our identity, that is, who we are in God. As we come to know ourselves in God and God in ourselves, we come to a deeper understanding of Jesus as the way of life. What takes place in Jesus is to take place in our lives as well, if we are to attain the fullness of life in God. As we saw with Bonaventure, spirituality is key to knowing Christ and the meaning of Christ crucified as the theological center, because all that can be known of God (theology) is revealed in the crucified Christ. Such knowledge is less of the mind than of the heart, for only the heart centered in God can know the truth of God and contemplate God, as God reveals Godself in fragile creation. The insights of our mystical guides indicate that spirituality is essential to doing Christology in the second axial period; prayer and grace must accompany intellectual strivings. Bede Griffiths reminds us that "it is only through the human being that the universe breaks through into a new reality."[3] Knowledge alone therefore is insufficient for the rebirth of Christ today. Rather, this new birth, like any new birth, must take place through crucified love, for love is borne in the surrender of the gift of self, that is, in relationship.

The deepening of our lives in Christ not only awakens us to who we are but leads to a new source of energy, a free, liberating energy that flows from the heart. Barnhart states that "awakening to transcendence or spiritual reality initiates a personal revolution by relativizing everything that one had previously experienced or known, opening a new depth of consciousness and a new relationship with this reality beyond the self."[4] One awakens to the realization that the meaning of Jesus the Christ is somehow the center of reality; one finds that one's life must be reordered to correspond to that realization. In Bonaventure's view, we begin to see (to contemplate) the mystery of Christ as the mystery of our lives. We are images of Christ; thus, we are created to bear Christ within us and to express the life of Christ in the world. The mystics indicate that to live the Christ mystery we must *participate* in the mystery. For Christians, such participation takes place in the sacramental life of the church and the sacramental life of the world where the church, the body of Christ, truly exists. The meaning of Christ must move from the altar of the church to the altar of the world. Participation in the world, integral to second-axial-period consciousness, is participation in Christ, insofar as belief in Christ empowers a new vision for the world which can unfold only in and through us. As we deepen our lives in Christ through an active spiritual life, we find the meaning of Christ "expanding" from the humanity of Jesus to the contemplation of Jesus in our brothers and sisters and the nonhuman creatures of earth. Without spirituality as the heart of Christology, however, we wind up in a sterile christomonism that stifles the meaning of Christ. Barnhart sums up this idea well when he writes,

> The christocentric intellectual revolution, when arrested, can lead to a vision which is not liberating but constricting. It can produce a "christomonism," reducing Christianity to the dimensions of a contained and containing Christ; that is, a Christ who has not been permitted to expand into his full magnitude, his full dimensionality. This is often a literalism or fundamentalism which wields the name of Christ as a blunt weapon, suppressing the dimensions of Father (or unitive depth) and of the Holy Spirit (or dynamism), insisting exclusively upon the objective image of Christ at the expense of the divinization of the person, the subjective realization of the Christ-event. The true christocentrism is a Copernican revolution in which every mode of thought, every philosophical and theological structure, must give way and find its orbital relation to the Word Incarnate and to the Christ-event as the living and active center of history.[5]

The significance of Christ in evolution is the complexification of the mystery of Christ in which the meaning of Christ has shifted from an individual, historical existent (Jesus of Nazareth) to the cosmic person symbolized by the humanity of Jesus. The emergence of individual consciousness and Jesus Christ as mediator of personal transcendence in the first axial period is taken up in the second axial period and transformed into a consciousness of relatedness. Every person of every religion and the natural world itself can now be understood in view of Christ. The emergence of globalization wrought by advances in science and technology, economics, and politics has ushered in a complexification of human consciousness and, in particular, religious consciousness. We are confronted today by interrelatedness, whether or not we accept our relatedness and find in it our identity. Unless we can begin to make metaphysical sense of religious pluralism, we run the risk of enhanced fundamentalism and fragmentation, as religions tightly secure their cultures and identities. Luiz Carlos Susin aptly describes the difficulties of the emerging paradigm:

> Faced with the pluralism and complexity of our age, contemporary societies are either exasperating each other by insisting on affirming their identities in religious fundamentalist accents or splitting into fragments with no unifying religious horizons ... Globalization can bring positive values, such as encouragement of democracy and human rights. Seen from the vast regions where the people are ever more excluded from its benefits, however, it forces peoples who feel its malevolent effects to seek ways of resisting in their humanity through recourse to the deepest sources of their religious roots. Religious pluralism can, then, appear ambiguous: on the one hand, it can be the fruit of fundamentalist resistance and even of violence in affirming one's own identity; on the other, it can be a legitimate and civilized expression of cultural identity, of its religious root, of its "soul" with its right to be different within human bio-diversity, resisting the overwhelming power of claims to universality based on the privilege of being the most powerful.[6]

There is no doubt that every evolutionary advance bears within it forces of resistance; every movement toward greater complexity struggles against a certain inertia. A problem may lie in the speed of evolutionary advance, which, as the specialists of technology indicate, is running faster than the human intellect can comprehend because of the ingenuity and creativity of humans themselves. Some may question whether or not the individual will be swallowed up in this new global tribe of humanity or in the noosphere of the Internet world. While fear may impel individuals to assert autonomy over and against one other, the true individual (and thus personality) will emerge through relatedness. Technology is enabling the discovery of a new type of personality through immediate communication and virtual reality. Thus, we may say that the second axial period bears with it a rediscovery of the individual as one who is communally related if not globally related, which underscores Teilhard's idea that in a christic universe things are themselves precisely in union with one another. My question is, however, as we advance in complexification of religious consciousness, where will the Christian stand, on the side of resistance or on the horizon of advance?

I have tried to show here that the emergence of global consciousness in the second axial period can help us realize that Christ is the organic, integrating center of creation. Life in Christ is not only ascending in God but it is also descending in God enfleshed, moving into solidarity with the other who is different from me but in whom God dwells. We can also now better apprehend the Christ-event not only on the scale of human history but on the scale of the evolutionary universe itself. Barnhart writes that "Jesus enters the evolutionary trajectory *omnem novitatem portans:* bringing all newness in himself, bringing the needed transformation, the consummation awaited by a dying (though ever evolving) world."[7] Through the human person a new reality emerges, born out of new structures of consciousness. Christ brings a "new heart" to humanity, both on the individual and on the collective plane. As Teilhard points out, humanity becomes a new "creative center" within the evolutionary process, in such a way that the path of this evolution now becomes explicitly directed; evolution has a goal. The "intelligent design" of creation is revealed. The evolutionary development "from cosmos to cosmogenesis" takes place under the influence of a divine creative power that has been present from the beginning but now made explicit in the human person. Christ becomes the concrete focal point, or Omega, of a new centration in the universe and Christian life, the growing tip of the whole evolutionary trend.

Panikkar uses the term "christophany" to liberate the title Christ from its "baptized" form and to posit the mystery of Christ as the central mystery of every person. As he writes, "Each person bears the mystery of Christ within. The first task of every creature, therefore, is to complete and perfect his or her icon of reality."[8] How each person perfects his or her icon of reality is integral to the mystery of Christ, as this mystery pertains to all world religions. Some will undoubtedly ask, What about those who do not know Christ or who do not believe in Christ? Do they participate in the evolution of Christ as well? Denis Edwards addresses this question when he writes,

> The human task of completing creation derives its meaning from the redemptive and divinizing will of God. This applies even to those who do not know the significance of their contributions. Those whose actions are directed toward the good of the cosmos, believers and unbelievers alike, fall under the impulse of grace. Their actions have eternal value.[9]

Whether or not one explicitly knows Jesus, the Christ mystery is at the core of human life and life in the universe. Because human nature is not accidental to Christ, Christ cannot be accidental to the human person; all of humanity/creation is intrinsic to the Christ mystery. Without us, Christ does not exist and without Christ we have no hope in the universe. Thus, every religion is ultimately centered in Christ insofar as every religion has a personal center and seeks unity in God. As Teilhard indicated, however, Christianity is the only religion that claims a personal and personalizing center of the universe.[10] Christians, therefore, bear a particular task in the universe, which is to help realize the fullness of Christ by following the example of Jesus. To be a Christian is to make Christ alive through relationships of love, to enter into dialogue with persons of other religions and cultures, to share life with others—to "put on" Christ. In Teilhard's view, this means plunging ourselves into the world, getting involved with the tribes of humanity, earth's people, and the earth itself. His "mysticism of action" includes the creative development of technology and realizes that human creativity and the emergence of technosapiens give new expression to Christ in evolution. Christ is the unifying mystery of what we are about with God if we are to have any hope in God.

Because we humans are in evolution we must see Christ in evolution as well—Christ's humanity is our humanity, Christ's life is our life. Although Christ is the energizing center and goal of creation, it is the Spirit continuously at work in creation who draws us into the mystery of Christ. The Spirit sent by Christ leads us to Christ and through Christ to the Father. That is why life in Christ can never be private or isolated, for Christ is the Word of the Father and the source of the Spirit. Christ is relational by definition and hence the source of community. To live Christ is to live community; to bear Christ in one's life is to become a source of healing love for the sake of community. Our guides for the second axial period open up to us this mystery of Christ and indicate that we must liberate Christ from a Western intellectual form that is logical, abstract, privatized, and individualized. We must engage in the complexification of Christ in the second axial period, which means accepting the diversity and differences of the other as integral to ourselves and thus integral to the meaning of Christ. Engagement with the other is not dissolving ourselves in the other but being true to ourselves—our identity—by finding ourselves in God and God in the other. Christ must be free *to be Christ* if this universe is to have any hope in God, and we must be free to be Christians if we are to be bearers of this hope.

Christ is the power of God among us and within us, the fullness of the earth and of life in the universe. We humans have the potential to make Christ alive; it is what we are created for. To live the mystery of Christ is not to speak about Christ but to live in the surrender of love, the poverty of being, and the cave of the heart. If we can allow the Spirit to really take hold of us and liberate us from our fears, anxieties, demands, and desire for power and control, then we can truly seek the living among the dead; we can live in the risen Christ who empowers us to build this new creation. We can look toward that time when there will be one cosmic person uniting all persons, one cosmic humanity uniting all humanity, one Christ in whom God will be all in all.

NOTES

Introduction

1. Cardinal Christoph Schönborn, "Finding Design in Nature," *New York Times*, July 7, 2005, at http://www.nytimes.com/2005/07/07/opinion/07schonborn.html.

2. www.catholicnews.com/data/stories/cns. See also Father George V. Coyne, "Science Does Not Need God. Or Does It? A Catholic Scientist Looks at Evolution." www.catholic.org. According to Coyne, evolution is not incompatible with Catholic doctrine because the scientific theory of evolution is completely neutral with respect to religious thinking. In his 1996 message John Paul II conceded that evolution is more than a hypothesis stating: "Today, more than a half-century after the appearance of that encyclical [*Humani Generis*], some new findings lead us toward the recognition of evolution as more than a hypothesis. In fact it is remarkable that this theory has had progressively greater influence on the spirit of researchers, following a series of discoveries in different scholarly disciplines. The convergence in the results of these independent studies—which was neither planned nor sought—constitutes in itself a significant argument in favor of the theory ... And to tell the truth, rather than speaking about the theory of evolution, it is more accurate to speak of the theories of evolution. The use of the plural is required here—in part because of the diversity of explanations regarding the mechanism of evolution, and in part because of the diversity of philosophies involved. There are materialist and reductionist theories, as well as spiritualist theories. Here the final judgment is within the competence of philosophy and, beyond that, of theology." See Pope John Paul II, "Message to the Pontifical Academy of Sciences: On Evolution." www.etwn.com/library/PAPALDOC.

3. John F. Haught, *Responses to 101 Questions on God and Evolution* (New York: Paulist Press, 2001), 5.

4. For a discussion of intelligent design, see William J. Dembski, *Intelligent Design: The Bridge between Science and Theology* (Westmont, Ill.: InterVarsity Press, 2002); Michael Behe, *Darwin's Black Box: The Biochemical Challenge to Evolution,* 2nd ed. (New York: Free Press, 2006).

For a summary account, see Paul Rasor, "Intelligent Design: Religion or Science?" *Religious Studies News* 22, no. 3 (May 2007): 19.

5. Catherine Baker and James Miller, eds., *The Evolution Dialogues: Science, Christianity, and the Quest for Understanding* (Washington, D.C.: American Association for the Advancement of Science, 2006), 158-63.

6. D. Bonhoeffer, *Letters and Papers from Prison* (New York: Macmillan, 1967), 139.

7. Sallie McFague, *Life Abundant: Rethinking Theology and Economy for a Planet in Peril* (Minneapolis: Fortress Press, 2001), 159.

8. Bernard McGinn, *The Presence of God: A History of Western Christian Mysticism* (New York: Crossroad, 1991), xvii.

9. Mary Meany, "Deconstruction and Reconstruction in Angela of Foligno," in *Divine Representations: Postmodernism and Spirituality*, ed. Ann W. Astell (New York: Paulist Press, 1994), 55.

10. *Angela of Foligno: Complete Works,* trans. and intro. Paul LaChance, preface by Romana Guernieri (New York: Paulist Press, 1993), 149.

11. *Angela of Foligno,* 169-70.

12. Margaret Carney, "Franciscan Women and the Theological Enterprise," in *The History of Franciscan Theology*, ed. Kenan B. Osborne (New York: Franciscan Institute, 1994), 336.

13. Bonaventure II *Sentences (Sent.)* d. 17, a. 1, q. 2, resp. (II, 414-15). The critical edition of Bonaventure's works used in this book is the *Opera Omnia,* ed. PP. Collegii S. Bonaventurae, 10 vols. (Quaracchi, 1882-1902). Latin texts are indicated by volume and page number in parentheses.

14. Bonaventure II *Sentences,* d. 12, a. 1, q. 3 concl. (II, 298); 195; Kent Emery, "Reading the World Rightly and Squarely: Bonaventure's Doctrine of the Cardinal Virtues," *Traditio* 39 (1983): 188-99. Bonaventure further indicates that God chose to effect this gradual perfection [of matter] over a course of six days because six, the sum and multiple of its integers, is the perfect number.

15. Zachary Hayes, "Christ, Word of God and Exemplar of Humanity," *Cord* 46, no. 1 (1996): 7.

16. Hayes, "Christ, Word of God and Exemplar of Humanity," 8.

17. Zachary Hayes, *A Window to the Divine: A Study of Christian Creation Theology* (Quincy, Ill.: Franciscan Press, 1997), 90.

18. Hayes, "Christ, Word of God and Exemplar of Humanity," 6.

19. Zachary Hayes, "The Meaning of *Convenientia* in the Metaphysics of St. Bonaventure," *Franciscan Studies* 34 (1974): 94.

20. Bonaventure III *Sentences,* d. 1, a. 2, q. 2, ad 5 (III, 26-27).

21. Hayes, "Christ, Word of God and Exemplar of Humanity," 14; Ilia Delio, "Revisiting the Franciscan Doctrine of Christ," *Theological Studies* 64, no. 1 (2003): 10.

22. Hayes, *Window to the Divine,* 91.

23. Hayes, "Christ, Word of God and Exemplar of Humanity," 13.

24. Bonaventure III *Sentences* d. 1, a. 2, q. 2, ad 6 (III, 27); Hayes, "The Meaning of *Convenientia*," 94.

25. Bonaventure III *Sentences,* d. 32, q. 5, ad 3 (III, 706).

26. Hayes, "Christ, Word of God and Exemplar of Humanity," 12.

27. Hayes, *Window to the Divine,* 91.

28. Hayes, "Christ, Word of God and Exemplar of Humanity," 12.

29. Hayes, "Christ, Word of God and Exemplar of Humanity," 13.

30. For a discussion of the Pauline notion of the cosmic Christ and its influence on the early church fathers, see George Maloney, *The Cosmic Christ: From Paul to Teilhard* (New York: Sheed & Ward, 1968).

31. Pierre Teilhard de Chardin, *Christianity and Evolution,* trans. René Hague (New York: Harcourt Brace Jovanovich, 1971), 87.

32. Raimon Panikkar, *Christophany: The Fullness of Man,* trans. Alfred DiLascia (Maryknoll, N.Y.: Orbis Books, 2004), 21.

33. Panikkar, preface to *Christophany,* p. xx.

34. William M. Thompson, *Jesus, Lord and Savior: A Theopatic Christology and Soteriology* (New York: Paulist Press, 1980), 256.

35. The taxonomy of animal and plant classification is typically broken down into kingdom, phylum, class, order, family, genus, and species (www.ecotao.com/holism/glosoz.htm). A phylum is the second largest group into which scientists classify living things. Each kingdom (Animal, Plant, Fungi, Protist, Moneran) is split into phyla. For example, the Chordate Phylum (animals with backbones) is in the Animal Kingdom. A phylum is a group of classes with shared characteristics. For example, the Chordate Phylum is made up of classes of animals that have spines or notochords.

36. Andrew J. Burgess, "Earth Chauvinism," *Christian Century* 93 (1976): 1098. Burgess defines exoChristology as Christological issues raised by discoveries in outer space.

1. Evolution, Christ, and Consciousness

1. Francisco J. Ayala, "Biological Evolution: An Introduction," in *An Evolving Dialogue: Theological and Scientific Perspectives on Evolution,* ed. James Miller (Harrisburg, Pa.: Trinity Press International, 2001), 10.

2. Ayala, "Biological Evolution," 10.

3. Ayala, "Biological Evolution," 13.

4. Robert John Russell, "Cosmology from Alpha to Omega," *Zygon* 29, no. 4 (December 1994): 557-56; Mark William Worthing, *God, Creation and Contemporary Physics* (Minneapolis: Fortress Press, 1996), 79-110.

5. David Toolan, *At Home in the Cosmos* (Maryknoll, N.Y.: Orbis Books, 2003),132-55; Simon Singh, *Big Bang: The Origin of the Universe* (New York: Harper, 2004), 144-61.

6. Singh, *Big Bang*, 116-43; Stephen W. Hawking, *A Brief History of Time* (New York: Bantam Books, 1998), 15-34; Worthing, *God, Creation and Physics*, 24-27.

7. John Polkinghorne, *Science and Theology* (Minneapolis: Fortress Press, 1998), 34; Hawking, *Brief History of Time*, 35-51.

8. For a summary of the work of Penzias, Wilson, and Gamow, see Stephen W. Hawking, *Brief History of Time*, 41- 43; Worthing, *God, Creation and Physics*, 28.

9. Ian Barbour, *Issues in Science and Religion* (London: SCM Press, 1997), 273-316; Hawking, *Brief History of Time*, 53-61.

10. Polkinghorne, *Science and Theology*, 39.

11. For a more detailed explanation of chaos theory, see Ilya Prigogene *Order out of Chaos* (reissue edition, New York: Bantam Books, 1984); Margaret Wheatley, *Leadership and the New Science: Learning about Organization from an Orderly Universe* (San Francisco: Berrett-Koehler Publishers, 1992), 75-99.

12. Polkinghorne, *Science and Theology*, 44.

13. Wheatley, *Leadership and the New Science*, 87-99.

14. Pierre Teilhard de Chardin, *Activation of Energy*, trans. René Hague (New York: Harcourt Brace Jovanovich, 1970), 387-403; idem, *Phenomenon of Man*, trans. Bernard Wall (New York: Harper & Row, 1959), 46-66.

15. Pierre Teilhard de Chardin, *Christianity and Evolution*, trans. René Hague (New York: Harcourt Brace Jovanovich, 1971), 87.

16. Elizabeth Johnson, *Women, Earth, and Creator Spirit* (New York: Paulist Press, 1993), 37.

17. John F. Haught, *God beyond Darwin: A Theology of Evolution* (Boulder, Colo.: Westview Press, 2000), 11-22.

18. Cameron Freeman, "An Evolutionary Christology," 5. www.flinders.edu.au/theology/evolution.

19. Ken Wilber, *Sex, Ecology, Spirituality: The Spirit of Evolution* (Boston, Mass.: Shambala Publications, 2000), 202-4.

20. See Robert Wright, *Non Zero: The Logic of Human Destiny* (New York: Vintage, 2001). Using language suggesting that natural selection is a designer's tool, Wright draws the conclusion that evolution is goal-oriented (or at least moves toward inevitable ends that are independent of environmental or contingent variables). He contends optimistically that history progresses in a predictable direction and points toward a certain end: a world of increasing human cooperation where greed and hatred have outlived their usefulness. This thesis is elaborated by way of something Wright calls "non-zero-sumness," which in game theory means a kind of win-win

situation. The non-zero-sum dynamic, Wright says, is the driving force that has shaped history from the very beginnings of life, giving rise to increasing social complexity, technological innovation, and, eventually, the Internet.

21. Daryl Dooming, "Evolution, Evil and Original Sin," *America* 185, no. 15 (November 12, 2001): 14-21; John Cobb (*The Structure of Existence* [New York: Seabury Press, 1979]) describes the appearance of the axial person as the emergence of freedom.

22. Freeman, "An Evolutionary Christology," 5-6.

23. Freeman, "An Evolutionary Christology," 6.

24. This is Teilhard de Chardin's basic thesis, as he describes it in his *Phenomenon of Man*.

25. Zachary Hayes, *A Window to the Divine: A Study of Christian Creation Theology* (Quincy, Ill.: Franciscan Press, 1997), 90.

26. Zachary Hayes, "Christology-Cosmology," in *Spirit and Life: A Journal of Contemporary Franciscanism*, vol. 7, ed. Anthony Carrozzo, Kenneth Himes, and Vincent Cushing (New York: Franciscan Institute, 1997), 43.

27. John F. Haught, *Christianity and Science: Towards a Theology of Nature* (Maryknoll, N.Y.: Orbis Books, 2007), xi.

28. Haught, *God beyond Darwin*, 5.

29. Hayes, *Window to the Divine*, 90.

30. For a discussion of noogenesis, see Teilhard de Chardin, *Phenomenon of Man*, 180-84.

31. Karl Jaspers, *The Origin and Goal of History*, trans. Michael Bullock (New Haven: Yale University Press, 1953), 1, 23, 27. While Jaspers's idea of an axial period has been helpful to understand the emergence of religious consciousness, it has also been identified as misconceived or rather a somewhat naïve notion of history. Eric Voegelin in his comprehensive work *The Ecumenic Age* criticizes Jaspers's axial period as one that ignored spiritual outbursts outside the time line proposed by Jaspers. Voegelin writes: "In order to elevate the period from 800 to 200 B.C. in which the parallel outbursts occur, to the rank of the great epoch in history, Jaspers had to deny to the earlier and later spiritual outbursts the epochal character which in their own consciousness they certainly had. In particular, he had to throw out Moses and Christ ... There was no 'axis time' in the first millennium B.C. because the Western and Far Eastern thinkers did not know of each others' existence and, consequently, had no consciousness of thinking on any axis of history. The 'axis time' I had to conclude, was the symbolism by which a modern thinker tried to cope with the disturbing problem of meaningful structures in history, such as the field of parallel spiritual movements, of which the actors in the field were quite unaware." Although Voegelin saw some merit in Jaspers's axial period, he also viewed it as a unilateral view of history and therefore of limited import. See Eric Voegelin, *The Ecumenic Age*, vol. 4, *Order and History* (Baton Rouge: Louisiana State University

Press, 1974), 4-5. In my view, the import of Jaspers's axial period should be treated as a heuristic notion, a *paradigm* of religious consciousness that we can now interpret within an evolutionary context. In this respect I favor John Cobb's appropriation of the axial period as integral to natural history which, he states, pertains to what is "universal, recurrent or widespread rather than to the specifics of history." Although Cobb recognizes that a number of features of Jaspers's axial period are "highly dubious," he sees great value in this concept for understanding the spiritual breakthroughs in the first millennium before Christ independently in China, India, Persia, Palestine, and Greece. Cobb speaks of these spiritual breakthroughs as "structures of existence" and sees the axial period as a plurality of structures of existence that proceeded side by side, as well as interacting, rather than speaking of a single new structure of existence expressing itself in several forms. It is in this context that Cobb discusses Christianity as a new structure of existence with a claim to finality. See John Cobb, *The Structure of Christian Existence* (New York: Seabury Press, 1979), 23-24.

32. Jaspers, *Origin and Goal of History*, 2.

33. William M. Thompson, *Christ and Consciousness: Exploring Christ's Contribution to Human Consciousness* (New York: Paulist Press, 1977), 21.

34. Ewert H. Cousins, *Christ of the 21st Century* (Rockport, Mass.: Element Books, 1992), 5.

35. Cobb, *Structure of Christian Existence*, 52- 59.

36. Thompson, *Christ and Consciousness*, 23.

37. Jaspers, *Origin and Goal of History*, 4.

38. Cobb, *Structure of Christian Existence*, 52-59; Thompson, *Christ and Consciousness*, 22-23.

39. Cousins, *Christ of the 21st Century*, 6; idem, "Teilhard's Concept of Religion and the Religious Phenomenon of Our Time," *Teilhard Studies* 49 (Fall 2004): 10-11.

40. Jaspers, *Origin and Goal of History*, 58.

41. Cousins, *Christ of the 21st Century*, 7.

42. Thompson, *Christ and Consciousness*, 22.

43. Cousins, "Teilhard's Concept of Religion," 11.

44. Thompson, *Christ and Consciousness*, 39.

45. Thompson, *Christ and Consciousness*, 63.

46. Thompson, *Christ and Consciousness*, 65.

47. Thompson, *Christ and Consciousness*, 67.

48. Gerald O'Collins, *The Resurrection of Jesus Christ* (Valley Forge, Pa.: Judson Press, 1973), 104.

49. Thompson, *Christ and Consciousness*, 67.

50. Cobb, *Structure of Christian Existence*, 122.

51. Cousins, *Christ of the 21st Century*, 7-10.

52. See Teilhard de Chardin, *Activation of Energy*, 30-31, 101-3; Cousins, "Teilhard's Concept of Religion," 12.

53. Cousins, "Teilhard's Concept of Religion," 13.

54. Justo Gonzalez, *Mañana: Christian Theology from a Hispanic Perspective* (Nashville: Abingdon Press, 1990), 110.

55. Teilhard de Chardin, *Christianity and Evolution*, 77.

56. Raimon Panikkar, "A Christophany for Our Time," *Theology Digest* 39, no. 1 (Spring 1992): 4.

57. Raimon Panikkar, *Christophany: The Fullness of Man*, trans. Alfred DiLascia (Maryknoll, N.Y.: Orbis Books, 2004), 5.

58. Panikkar, *Christophany*, 5.

59. Panikkar, "A Christophany for Our Time," 8.

60. Panikkar, "A Christophany for Our Time," 8; idem, *Christophany*, 150.

61. Panikkar, *Christophany*, 5.

62. Teilhard de Chardin, *Christianity and Evolution*, 77.

63. Teilhard de Chardin, *Christianity and Evolution*, 89.

64. Teilhard de Chardin, *Christianity and Evolution*, 78.

2. A Brief History of Christ

1. This idea is the basis of systems theory in which the idea of independent parts connected to one another has given way to interdependent parts related to one another. See, e.g., Fritjof Capra, *The Web of Life: A New Understanding of Living Systems* (New York: Anchor Books, 1997), 35-49.

2. N. Max Wildiers, *The Theologian and His Universe: Theology and Cosmology from the Middle Ages to the Present*, trans. Paul Dunphy (New York: Seabury Press, 1982), 38.

3. John L. McKenzie, *Dictionary of the Bible* (New York: Macmillan, 1965), 434; Ben F. Meyer, "Jesus Christ," *The Anchor Bible Dictionary* (New York: Doubleday, 1992), 3:773.

4. Christopher Tuckett, *Christology and the New Testament: Jesus and His Earliest Followers* (Louisville, Ky.: Westminster John Knox Press, 2001), 16. According to Joseph Fitzmyer, the Jews had various messianic expectations at the time of Jesus. Some Jews expected a royal messiah, others a priestly messiah, others saw the messiah as both royal and priestly, and some Jews had no fervent expectation of a coming messiah. See Joseph Fitzmyer, *Christological Catechism: New Testament Answers* (New York: Paulist Press, 1991).

5. Tuckett, *Christology and the New Testament*, 16-19. Tuckett indicates that messianic language among the Jews was associated with either a royal figure or a priestly figure who would help inaugurate the new age. The

messianic ideas taken up and applied to Jesus seem to be royal ideas. Tuckett states, "One of the great conundrums of the study of the New Testament Christology is how and why this happened: for Jesus' whole life and work are not clearly related to ideas associated with a royal figure; and yet the term *christos*/Messiah is already attached to Jesus very firmly and very early so that it soon becomes just another proper name as in 'Jesus Christ' or 'Christ Jesus' or even 'Christ' (cf. 1 Cor 15:3)" (p. 19). See also Joseph Fitzmyer, *The One Who Is to Come* (Grand Rapids: Eerdmans, 2006).

6. Jürgen Moltmann, *The Way of Jesus Christ: Christology in Messianic Dimensions*, trans. Margaret Kohl (New York: HarperCollins, 1990), 73.

7. Moltmann, *Way of Jesus Christ*, 91.

8. William M. Thompson, *Jesus, Lord and Savior: A Theopatic Christology and Soteriology* (New York: Paulist Press, 1980), 174.

9. Albert Schweitzer, *The Quest of the Historical Jesus: A Critical Study of Its Progress from Reimarus to Wrede*, trans. W. Montgomery (New York: Macmillan, 1910, 1960), 330-403. It should be noted that Schweitzer's position has recently been challenged. Marcus Borg states that "a major paradigm shift away from an eschatological understanding of Jesus' teaching may be occurring, even though no replacement consensus has emerged." See Marcus Borg, "The Teaching of Jesus Christ," *The Anchor Bible Dictionary* (New York: Doubleday, 1992), 3:806.

10. Thompson, *Jesus, Lord and Savior*, 177.

11. Thompson notes that the word *persona* comes from both *per* and *sonare*, "to sound through." One becomes a self/person through allowing both the vertical and horizontal dimensions to sound through one. See Thompson, *Jesus, Lord and Savior*, 169. For a discussion of the meaning of "person," see Joseph W. Koterski, "Boethius and the Theological Origins of the Concept of Person," *American Catholic Philosophical Quarterly* 78, no. 2 (2004): 205-6.

12. Reginald H. Fuller, *A Critical Introduction to the New Testament* (London: Duckworth, 1966), 109.

13. Fuller, *Critical Introduction*, 110.

14. John L. McKenzie, *Dictionary of the Bible* (New York: Macmillan, 1965), 434; Ben F. Meyer, "Jesus Christ," *The Anchor Bible Dictionary*, (New York: Doubleday, 1992), 3:787.

15. Tuckett (*Christology and the New Testament*, 46) writes: "The title Messiah essentially referred to a figure of Jewish eschatological expectation. This figure was thought of as a royal figure, a priestly figure or a prophetic figure, with the royal idea probably the dominant one." The association of Jesus' messiahship with death and resurrection completely overturned Jewish expectations, a "stumbling block for the Jews," as Paul writes (1 Cor 1:23).

16. N. T. Wright, "The Historical Jesus and Christian Theology"

www.ntwrightpage.com/Wright. See also N. T. Wright, *Jesus and the Victory of God*, vol. 2, *Christian Origins and the Question of God* (Minneapolis: Fortress Press, 1996), 477-86. It is noted here that Wright holds a minority position on Jesus' messiahship that has not been entirely refuted and thus holds interest.

17. N. T. Wright, "How Jesus Saw Himself," *Bible Review* 12, no. 3 (1996): 27.

18. Wright, "How Jesus Saw Himself," 28; Wright, *Jesus and the Victory of God*, 413-15.

19. Wright, "How Jesus Saw Himself," 28-29. See also Marcus J. Borg and N. T. Wright, *The Meaning of Jesus: Two Visions* (New York: Harper-Collins, 1999), 164-66. Wright offers a convincing argument that the early disciples of Jesus recognized him as Messiah. That is, Jesus as the Christ is not entirely contingent on the resurrection; rather, his life portrayed messianic meaning, which Wright identifies in the symbolic action in the Temple (Luke 14:27-32) and cryptically in a series of riddles. See Wright, *Jesus and the Victory of God*, 486-519. Wright states that "Jesus saw his work from the start as in some sense messianic" (p. 532).

20. Wright, "The Historical Jesus and Christian Theology."

21. Cited in Paul F. Knitter, *No Other Name? A Critical Survey of Christian Attitudes Toward the World Religions* (Maryknoll, N.Y.: Orbis Books, 1985), 179.

22. James D. G. Dunn, *Christology in the Making* (Philadelphia: Westminster, 1980), 265-67; see also p. 62. It is interesting to note that the early Christians (ca. A.D. 30-50) held an exalted view of Jesus and treated him as a recipient of religious devotion and associated him with God in striking ways. See Larry Hurtado, *Lord Jesus Christ: Devotion to Jesus in Earliest Christianity* (Grand Rapids: Eerdmans, 2003), 2.

23. Wright, "The Historical Jesus and Christian Theology"; Wright, *Jesus and the Victory of God*, 658.

24. Kenan B. Osborne, *The Resurrection of Jesus: New Considerations for Its Theological Interpretation* (New York: Paulist Press, 1997), 117.

25. Roger Haight, *Jesus Symbol of God* (Maryknoll, N.Y.: Orbis Books, 1999), 123.

26. William M. Thompson, "The Risen Christ, Transcultural Consciousness, and the Encounter of the World Religions," *Theological Studies* 37 (1976): 403.

27. Gerald O'Collins, *The Resurrection of Jesus Christ* (Valley Forge, Pa.: Judson Press, 1973), 52. However, as N.T. Wright points out, when the early Christians said "resurrection" they meant bodily resurrection and nothing other. See N. T. Wright, *The Resurrection of the Son of God*, vol. 3, *Christian Origins and the Question of God* (Minneapolis: Fortress Press, 2003), 209.

28. Historical study of the resurrection is complex and the reader is referred to the magisterial work of Wright (*Resurrection of the Son of God*) as well as the recent book by Fitzmyer (*The One Who Is to Come*) on the meaning of the term "messiah" as an eschatological figure, an anointed human agent of God, sent by God and awaited at the end of time.

29. William M. Thompson, *Christ and Consciousness: Exploring Christ's Contribution to Human Consciousness* (New York: Paulist Press, 1977), 83 n. 28.

30. Haight, *Jesus Symbol of God*, 124.

31. Cletus Wessels, *Jesus in the New Universe Story* (Maryknoll, N.Y.: Orbis Books, 2003), 132.

32. Thompson, *Christ and Consciousness*, 66-80.

33. Actually Thompson interprets Merton's transcultural Christ as the "finally integrated man." See Thompson, *Jesus, Lord and Savior*, 262.

34. Thompson, "The Risen Christ," 405.

35. Bonaventure, *Sermo I, Dom. II in Quad.* (IX, 215-19). Engl. trans., Zachary Hayes, "Christ, Word of God and Exemplar of Humanity," *Cord* 46, no. 1 (1996): 13.

36. Tuckett, *Christology and the New Testament*, 70; James D. G. Dunn, *The Theology of Paul the Apostle* (Grand Rapids: Eerdmans, 1997).

37. Tuckett, *Christology and the New Testament*, 71.

38. It is not certain that these letters were written during Paul's Roman imprisonment, but this is the traditional view. See Fuller, *Critical Introduction to the New Testament*, 31, 38.

39. Jürgen Becker, *Paul: Apostle to the Gentiles*, trans. O. C. Dean (Louisville, Ky.: Westminster John Knox Press, 1993), 317; Fuller, *Critical Introduction to New Testament*, 34.

40. According to Becker (*Paul: Apostle to the Gentiles*), assertions of the theology of the cross do not occur in the first early Christian generation except in Paul (p. 317).

41. Becker, *Paul*, 317-18.

42. Tuckett, *Christology and the New Testament*, 72.

43. J. Christiaan Beker, *Heirs of Paul: Paul's Legacy in the New Testament and in the Church Today* (Minneapolis: Fortress Press, 1991), 66.

44. Moltmann, *Way of Jesus Christ*, 255.

45. Tuckett, *Christology and the New Testament*, 76.

46. Tuckett, *Christology and the New Testament*, 77.

47. Moltmann, *Way of Jesus Christ*, 280.

48. Tuckett, *Christology and the New Testament*, 77.

49. Tuckett, *Christology and the New Testament*, 77.

50. Michael D. Meilach, *The Primacy of Christ in Doctrine and Life* (Chicago: Franciscan Herald Press, 1964), 71. For an extensive treatment of the Pauline literature and the primacy of Christ, see Jean-François Bon-

nefoy, *Christ and the Cosmos*, trans. Michael Meilach (Paterson, N.J.: St. Anthony Guild, 1965).

51. Tuckett, *Christology and the New Testament*, 78-79.

52. Tuckett, *Christology and the New Testament*, 82.

53. Tuckett, *Christology and the New Testament*, 82.

54. See George Maloney, *The Cosmic Christ: From Paul to Teilhard* (New York: Sheed & Ward, 1968), 30-36.

55. Maloney, *Cosmic Christ*, 15.

56. Irenaeus of Lyons 3.16.6 (*PG* 7.925-926), cited in Meilach, *Primacy of Christ*, 6.

57. Basil Studer, *Trinity and Incarnation: The Faith of the Early Church*, trans. Matthias Westerhoff; ed. Andrew Louth (Collegeville, Minn.: Liturgical Press, 1993), 109.

58. Athanasius, *De incarnatione* 54. Engl. transl. *Athanasius: Contra gentes and De incarnatione*, trans. and ed. Robert W. Thomson (Oxford: Clarendon Press, 1971), 269.

59. John Macquarrie, *Christology Revisited* (Harrisburg, Pa.: Trinity Press International, 1998), 45.

60. R. V. Sellers, *The Council of Chalcedon* (London: SPCK, 1961), 203.

61. Sellers, *Council of Chalcedon*, 210-11.

62. Aloys Grillmeier, *Christ in the Christian Tradition*, vol. 1, *From the Apostolic Age to Chalcedon (451)*, trans. John Bowden, revised ed. (Atlanta: John Knox Press, 1975), 545.

63. *The Christological Controversy*, trans. and ed. Richard A. Norris (Philadelphia: Fortress Press, 1980), 31.

64. Moltmann, *Way of Jesus Christ*, 274.

65. J. A. Lyons, *The Cosmic Christ in Origen and Teilhard de Chardin* (London: Oxford University Press, 1982), 7.

66. Lyons, *Cosmic Christ*, 9-10. Wildiers (*The Theologian and His Universe*, 7) discusses the semantic development of the Greek term *kosmos*, stating, "The term cosmos was initially employed to denote order, organization, regulation and can refer to all areas of life. To act *Kata kosmon* means to act in the right way, to behave properly. When referring to things or objects it means that these be designed correctly and be useful and efficient. In reference to people, however, it means that they live in an orderly manner such as is the case in an army or in the Greek city-state. In this way, the word cosmos initially belongs entirely and exclusively to the human social world and has as its counterpart the word *akosmia*, meaning disorder or chaos. The term cosmos was gradually used to designate the universe, thereby acquiring a technical meaning it did not previously have. This shift in meaning clearly dates from the fourth century B.C., as can be seen from texts of Plato."

67. Lyons, *Cosmic Christ*, 1-2.

68. Moltmann, *Way of Jesus Christ*, 276; Lyons, *Cosmic Christ*, 59.

69. Cited in Lyons, *Cosmic Christ*, 59.

70. Cited in Moltmann, *Way of Jesus Christ*, 277.

71. Joseph A. Sittler, "Called to Unity," *The Ecumenical Review* 14 (1961): 184-86.

72. Sittler, "Called to Unity," 178.

73. Sittler, "Called to Unity," 183-84, 187.

74. Moltmann, *Way of Jesus Christ*, 277.

75. Moltmann, *Way of Jesus Christ*, 279.

3. Franciscan Cosmic Christology

1. Michael D. Meilach, *The Primacy of Christ in Doctrine and Life* (Chicago: Franciscan Herald Press, 1964), 7.

2. Meilach, *Primacy of Christ*, 7.

3. Ilia Delio, "Revisiting the Franciscan Doctrine of Christ," *Theological Studies* 63 (2003): 3-4; Bernard McGinn, "Christ as Savior in the West," in *Christian Spirituality: Origins to the Twelfth Century*, ed. Bernard McGinn and John Meyendorff, *World Spirituality: An Encyclopedic History of the Religious Quest*, vol. 16 (New York: Crossroad, 1987), 256; Meilach, *Primacy of Christ*, 8.

4. Thomas Aquinas, *Summa Theologica* 3.1.3 *Opera Omnia*, vol. 11 (Rome: Vatican, 1903), 13-14.

5. For a discussion of the cosmic Christology of the Greek fathers, see George Maloney, *The Cosmic Christ: From Paul to Teilhard* (New York: Sheed & Ward, 1968), 142-81; and on the two principal medieval positions for the incarnation, see Meilach, *Primacy of Christ*, 27-58. Most theologians use the terms *motive* and *reason* interchangeably when referring to God's purpose in becoming incarnate. J. M. Bonnefoy, however, objected strongly to the former term because he felt it involved the connotation of influence by an outside agent. He maintained that since nothing outside God can influence him to act, the term *reason* is more appropriate in this context. See Meilach, *Primacy of Christ*, 199 n. 41.

6. The work of the Franciscan scholar Dominic Unger has done much to further our understanding of the primacy of Christ through the writings of the fathers of the church. See D. J. Unger, "Christ Jesus, the Center and Final Scope of All Creation according to St. Maximus the Confessor," *Franciscan Studies* 9 (1949): 50-62; idem, "Christ Jesus, the Exemplar and Final Scope of All Creation according to Anastasius of Sinai," *Franciscan Studies* 9 (1949): 156-64; idem, "Christ Jesus, the Secure Founda-

tion according to St. Cyril of Alexandria," *Franciscan Studies* 7 (1947): 1-25, 324-43; idem, "The Incarnation, a Supreme Exaltation for Christ according to St. John Damascene," *Franciscan Studies* 8 (1948): 247-49; idem, "The Love of God as the Primary Reason for the Incarnation according to Isaac of Niniveh," *Franciscan Studies* 9 (1949): 146-55.

7. Alexander of Hales, *Quaestiones disputatae: "Antequam esset frater,"* (Quarrachi: Collegium S. Bonaventurae, 1960), 197.

8. Kenan B. Osborne, "Alexander of Hales," in *The History of Franciscan Theology*, ed. Kenan B. Osborne (New York: Franciscan Institute, 1994), 31.

9. Osborne, "Alexander of Hales," 31.

10. Zachary Hayes, "Incarnation and Creation in the Theology of St. Bonaventure," in *Studies Honoring Ignatius Brady, Friar Minor*, ed. Romano Stephen Almagno and Conrad Harkins (New York: Franciscan Institute, 1976), 315.

11. Zachary Hayes, "Christ, Word of God and Exemplar of Humanity," *Cord* 46, no. 1 (1996): 6.

12. Hayes, "Christ, Word of God and Exemplar of Humanity,"6.

13. This quotation is found in the *Opus Oxoniensis* Ms 137, fol. 174b in the Biblioteca de Assisi: "Volendo alios condiligere idem obiect secum … Quare primo se amat ordinate et per consequens non inordinate, zelando vel invidendo, secundo vult alios habere condiligentes." Cited in Giovanni Iammarrone, "The Timeliness and Limitations of the Christology of John Duns Scotus for the Development of a Contemporary Theology of Christ," trans. Ignatius McCormick, *Greyfriars Review* 7, no. 2 (1993): 233 n. 13.

14. Allan B. Wolter, *Duns Scotus: Four Questions on Mary* (intro., text, and trans.) (Santa Barbara, Calif.: Old Mission Santa Barbara, 1988), 29.

15. Mary Beth Ingham, "John Duns Scotus: An Integrated Vision," in *The History of Franciscan Theology*, ed. Osborne, 221.

16. Allan B. Wolter, "John Duns Scotus on the Primacy and Personality of Christ," in *Franciscan Christology*, ed. Damian McElrath, Franciscan Sources, no. 1, ed. George Marcil (New York: Franciscan Institute, 1980), 141.

17. Alexander Gerken, *La théologie de verbe: La relation entre l'incarnation et la creation selon S. Bonaventure* (Paris: Éditions Franciscaines, 1970), 309-19.

18. Bonaventure I *Sentence* (*Sentences*). d. 27, p.1, a. un., q. 2, ad 3 (I, 470). The idea that the Father is innascible and fecund underlies the dialectical style of Bonaventure's thought. It also provides the basis of Bonaventure's metaphysics as a *coincidentia oppositorum*. The Father's innascibility and fecundity are mutually complementary opposites that cannot be formally reduced to one or the other; the Father is generative precisely

because he is unbegotten. See Zachary Hayes, introduction to *Disputed Questions on the Mystery of the Trinity,* vol. 3, *Works of Saint Bonaventure,* ed. George Marcil (New York: Franciscan Institute, 1979), 42, n. 51.

19. Bonaventure I *Sentences* d. 5, a. 1, q. 2, resp. (I, 115); I *Sentences* d. 2, a. u., q. 4, fund 2 (I, 56); Hayes, introduction, 34, n. 10. Bonaventure uses the terms *per modum naturae* and *per modum voluntatis* to designate the two trinitarian emanations. The terms are inspired by Aristotle's principle that there exist only two perfect modes of production; namely, natural and free.

20. Bonaventure I *Sentences* d. 6, a.u., q. 2, resp. (I, 128). "Processus per modum voluntatis concomitante natura"; Kevin P. Keane, "Why Creation? Bonaventure and Thomas Aquinas on God as Creative Good," *Downside Review* 93 (1975): 15. Keane writes, "It is noteworthy that Bonaventure's reason for attributing creation to the divine will is quite different from Thomas's. Where Thomas is in the main concerned to protect the divine perfection and radically free will, Bonaventure is at pains to elucidate how only through the will can an act be truly personal—both free and expressive of the outward dynamism of goodness, an act spontaneous yet substantial."

21. Hayes, "Incarnation and Creation in St. Bonaventure," 314.

22. Hayes, "Incarnation and Creation in St. Bonaventure," 314.

23. Zachary Hayes, "Christology and Metaphysics in the Thought of Bonaventure," *Journal of Religion* 58 (1978): S91.

24. Hayes, "Incarnation and Creation in St. Bonaventure," 314.

25. Hayes, "Incarnation and Creation in St. Bonaventure," 311.

26. Hayes, "Incarnation and Creation in St. Bonaventure," 311.

27. Bonaventure III *Sentences* d. 1, a. 1, q. 1, ad 1 (III, 10) ; ad 3 (III, 10-11) ;III *Sentences* d. 1, a. 1, q. 2, ad 2 (III, 13); ad 4 (III, 13); ad 6 (III, 13).

28. Zachary Hayes, "The Meaning of *Convenientia* in the Metaphysics of St. Bonaventure," *Franciscan Studies* 34 (1974): 78. Hayes notes the conditions for the possibility of incarnation by stating: (1) the two terms of the relation must be capable of entering into such a unique and intense union; (2) there must be a unity of person; for if this were not the case, then the history of Jesus would not be the history of the Word but a history only extrinsically related to the Word; and (3) granted the possibility from the side of God and from the side of man, it is yet required that there be a power adequate to effect the union.

29. Hayes, "Meaning of *Convenientia,*" 77.

30. Bonaventure, *Collationes in Hexaëmeron* (*Hex.*) 9, 2 (V, 373); 3, 7 (V, 344).

31. Zachary Hayes, *The Hidden Center: Spirituality and Speculative Christology in St. Bonaventure* (New York: Franciscan Institute, 1992), 60.

32. Hayes, "Christ, Word of God and Exemplar of Humanity," 14; Delio, "Revisiting the Franciscan Doctrine of Christ," 10.

33. Hayes, "Christ, Word of God and Exemplar of Humanity," 10.

34. Hayes, "Christ, Word of God and Exemplar of Humanity," 6. For a discussion on Bonaventure's soteriology and the primacy of Christ, see Zachary Hayes, *The Hidden Center*, 152-91, especially 189-91.

35. Hayes, "Meaning of *Convenientia*," 94.

36. Bonaventure III *Sentences* d. 1, a. 2, q. 2, ad 5 (III, 26-27).

37. Hayes, "Christ, Word of God and Exemplar of Humanity," 14; Delio, "Revisiting the Franciscan Doctrine of Christ," 10.

38. Zachary Hayes, *Bonaventure: Mystical Writings* (New York: Crossroad, 1999), 112.

39. Bonaventure, *Itinerarium Mentis in Deum* (Itin.) 2.1 (V, 300).

40. Bonaventure, *Breviloquium* (*Brev.*) 2.12 (V, 230).

41. Bonaventure, *Breviloquium* 2.12 (V, 230). Engl. trans. Dominic V. Monti, *Breviloquium*, vol. 9, *Works of St. Bonaventure*, ed. Robert J. Karris (New York: Francisan Institute Publications, 2005), 96.

42. Denis Edwards, "The Discovery of Chaos and the Retrieval of the Trinity," in *Chaos and Complexity: Scientific Perspectives on Divine Action* (Rome: Vatican Observatory Publications, 1995), 162.

43. Edwards, "Discovery of Chaos," 163.

44. See Mary Beth Ingham, *Scotus for Dunces: A Simple Guide to the Subtle Doctor* (New York: Franciscan Institute, 2003), 53-54.

45. Karl Rahner, *Theological Investigations*, vol. 5, *Later Writings*, trans. Karl-H. Kruger (Baltimore: Helicon Press, 1961), 164; idem, *Foundations of Christian Faith: An Introduction to the Idea of Christianity*, trans. W. V. Dych (New York: Seabury Press, 1978), 184.

46. Rahner, *Later Writings*, 179; idem, *Foundations*, 108.

47. Rahner, *Later Writings*, 163; idem, *Foundations*, 183.

48. Hayes, "Christ, Word of God and Exemplar of Humanity," 12.

49. Zachary Hayes, *A Window to the Divine: A Study of Christian Creation Theology* (Quincy, Ill.: Franciscan Press, 1997), 91.

50. Hayes, *Window to the Divine*, 90.

51. In the introduction to his book *Science and Religion: From Conflict to Conversation*, John Haught describes the confession of a scientific skeptic who maintains there is nothing beyond the physical universe. According to this skeptic, "There is not a shred of evidence that the universe is purposeful or that it is influenced by any kind of deity who cares for me ... As for why we are here, there is no other explanation than sheer chance ... After becoming familiar with neo-Darwinian theories of evolution it is hard to imagine how any intelligent person can adhere to the idea of a purposeful universe." See John F. Haught, *Science and Religion: From Conflict to Conversation* (New York: Paulist Press, 1995), 6-7.

52. Hayes, "Christ, Word of God and Exemplar of Humanity," 8.

53. Hayes, "Christ, Word of God and Exemplar of Humanity," 12.

54. Hayes, "Christ, Word of God and Exemplar of Humanity," 13.

55. Hayes, *Window to the Divine*, 91.

56. Rahner, *Foundations*, 201. One can begin to see here the compatibility between Rahner and Bonaventure's thought on the significance of Christ.

57. Rahner, *Foundations*, 284.

58. John Macmurray, *Persons in Relation* (London: Faber & Faber, 1961). Cited in William A. Barry, "God's Sorrow: Another Source of Resistance?" *Review for Religious* 48, no. 6 (1989): 843.

59. Hayes, "Christ, Word of God and Exemplar of Humanity," 13.

60. Bonaventure, *De reductione artium ad theologiam* (*Red. art.*) 20 (V, 324b).

4. Teilhard de Chardin: The Christic Universe

1. Jürgen Moltmann, *The Way of Jesus Christ: Christology in Messianic Dimensions*, trans. Margaret Kohl (New York: HarperCollins, 1990), 41.

2. Moltmann, *Way of Jesus Christ*, 69.

3. Moltmann, *Way of Jesus Christ*, 69.

4. Bernard McGinn, "The Venture of Mysticism in the New Millennium," Sophia Lecture, Washington Theological Union, D.C., February 11, 2007, p. 16.

5. N. Max Wildiers, *The Theologian and His Universe: Theology and Cosmology from the Middle Ages to the Present*, trans. Paul Dunphy (New York: Seabury Press, 1982), 205-6.

6. Thomas M. King, *Teilhard's Mysticism of Knowing* (New York: Seabury Press, 1981), 32.

7. King, *Mysticism of Knowing*, 32.

8. Pierre Teilhard de Chardin, *Phenomenon of Man*, trans. Bernard Wall (New York: Harper & Row, 1959), 110.

9. Pierre Teilhard de Chardin, *Activation of Energy*, trans. René Hague (New York: Harcourt Brace Jovanovich, 1970), 124.

10. King, *Mysticism of Knowing*, 33.

11. Teilhard de Chardin, *Phenomenon of Man*, 165.

12. Teilhard de Chardin, *Phenomenon of Man*, 221.

13. Pierre Teilhard de Chardin, *Human Energy*, trans. J. M. Cohen (New York: Harcourt Brace Jovanovich, 1969), 23.

14. King, *Mysticism of Knowing*, 34.

15. Robert Faricy, *The Spirituality of Teilhard de Chardin* (Minneapolis: Winston Press, 1981), 32-33.

16. Teilhard de Chardin, *The Phenomenon of Man*, 297-98.

17. Pierre Teilhard de Chardin, *Christianity and Evolution*, trans. René Hague (New York: Harcourt Brace Jovanovich, 1971), 87.

18. Teilhard de Chardin, *Christianity and Evolution*, 87.

19. Teilhard de Chardin, *The Phenomenon of Man*, 262.

20. Teilhard de Chardin, *The Phenomenon of Man*, 271.

21. Teilhard de Chardin, *Christianity and Evolution*, 87-88.

22. Teilhard de Chardin, *Phenomenon of Man*, 293-94; Timothy Jamison, "The Personalized Universe of Teilhard de Chardin," in *There Shall Be One Christ*, ed. Michael Meilach (New York: Franciscan Institute, 1968), 26.

23. Cited in Henri de Lubac, *Teilhard de Chardin: The Man and His Meaning*, trans. René Hague (New York: Hawthorn Books, 1965), 32.

24. Teilhard de Chardin, *Phenomenon of Man*, 297; de Lubac, *Teilhard de Chardin*, 32-33.

25. Teilhard de Chardin, *Christianity and Evolution*, 181; de Lubac, *Teilhard de Chardin*, 37.

26. Pierre Teilhard de Chardin, *My Universe*, trans. René Hague in *Process Reality*, ed. Ewert Cousins (New York: Newman Press, 1971), 251.

27. De Lubac, *Teilhard de Chardin*, 37.

28. De Lubac, *Teilhard de Chardin*, 42. Teilhard uses the mathematical description of an indefinite number, "n," to describe the number of possible planets with intelligible life.

29. Teilhard de Chardin, *Christianity and Evolution*, 176.

30. Teilhard de Chardin, *Christianity and Evolution*, 181.

31. De Lubac, *Teilhard de Chardin*, 47.

32. Ursula King, "'Consumed by Fire from Within': Teilhard de Chardin's Pan-christic Mysticism in Relation to the Catholic Tradition," *Heythrop Journal* 40, no. 4 (1999): 456-77.

33. Letter of February 9, 1916, to Berthe Teilhard de Chardin, quoted in Pierre Teilhard de Chardin, *The Heart of Matter*, trans. René Hague (New York: Harcourt Brace Jovanich, 1979), 188.

34. Unpublished retreat notes 1939-1943, quoted in Faricy, *The Spirituality of Teilhard de Chardin*, 17.

35. Faricy, *Spirituality of Teilhard de Chardin*, 22-23.

36. See Denis Edwards, *Jesus and the Cosmos* (New York: Paulist Press, 1991), 99-108.

37. Zachary Hayes, "Christ, Word of God and Exemplar of Humanity," *Cord* 46, no. 1 (1996): 7.

38. Teilhard de Chardin, *Christianity and Evolution*, 179; J. A. Lyons, *The Cosmic Christ in Origen and Teilhard de Chardin* (London: Oxford University Press, 1982), 183-96. The concept of a "third nature" is difficult to grasp if we conceive of nature as substantial being. Although Teil-

hard's insight is more mystical than scientific, we can describe this third nature along the lines of what science is telling us today about matter, namely, it is thoroughly relational. There is no part that is not related to other parts and every part is the result of its relationships. Hence, the whole is in every part and every part represents the whole. If Christ is the divine Word truly incarnate, then every aspect of matter is truly Christ and Christ is every aspect of matter. The humanity of Christ symbolizes not only the meaning of Christ for humanity but for the entire creation. In short, the cosmic or third nature that Teilhard describes is not a figurative description but a literal one. It underlies his Mass on the World in which he could offer up all his activities and all creation "through him, with him, and in him" to the glory of the Father.

39. Teilhard de Chardin, *Christianity and Evolution*, 181.

40. Lyons, *Cosmic Christ in Origen and Teilhard de Chardin*, 185-86.

41. Bonaventure, *Sermo I, Dom. II in Quad.* (IX, 215-19). Engl. trans. Hayes, "Christ, Word of God and Exemplar of Humanity," 13.

42. De Lubac, *Teilhard de Chardin*, 49.

43. Teilhard de Chardin, *Christianity and Evolution*, 89.

44. Teilhard de Chardin, *Christianity and Evolution*, 88.

45. Teilhard de Chardin, *Christianity and Evolution*, 89.

46. Teilhard de Chardin, *Christianity and Evolution*, 91-92.

47. Teilhard de Chardin, *Christianity and Evolution*, 92.

48. Teilhard de Chardin, *Christianity and Evolution*, 76-77.

49. Zachary Hayes, "Christology-Cosmology," in *Spirit and Life: A Journal of Contemporary Franciscanism*, vol. 7, ed. Anthony Carrozzo, Kenneth Himes, and Vincent Cushing (New York: Franciscan Institute, 1997), 44.

50. Teilhard de Chardin, *Christianity and Evolution*, 77.

51. Teilhard de Chardin, *Christianity and Evolution*, 78.

52. Ursula King, *Christ in All Things* (Maryknoll, N.Y.: Orbis Books, 1997), 80.

53. King, *Teilhard's Mysticism of Knowing*, 37.

54. George M. Maloney, *The Cosmic Christ: From Paul to Teilhard* (New York: Sheed and Ward, 1968), 184-85.

55. According to Claude Tresmontant, Teilhard followed Franciscan thought regarding the incarnation and redemption. See Claude Tresmontant, *Pierre Teilhard de Chardin*, trans. Salvator Attanasio (Baltimore: Helicon Press, 1959), 97.

56. Bonaventure, *Itinerarium Mentis in Deum* 1.9-15 (V, 299).

57. Ewert Cousins, "Teilhard de Chardin and Saint Bonaventure," in *There Shall Be One Christ*, ed. Meilach, 11-12.

58. Cousins, "Teilhard de Chardin and Saint Bonaventure," 17.

59. Teilhard de Chardin, *Phenomenon of Man*, 264.

60. Pierre Teilhard de Chardin, *La parole attendue* in *Cahiers,* vol. 4 (Paris: Editions du Seuil, 1963), 27.

61. Cousins, "Teilhard de Chardin and Saint Bonaventure," 17.

62. Maloney, *Cosmic Christ,* 211.

63. Maloney, *Cosmic Christ,* 189; Wildiers, *The Theologian and His Universe,* 207. Wildiers writes: "The completion of the world in Christ is not imposed on us as a necessity but is offered to us as a possibility that will not be realized without our cooperation. The further evolution of humankind ought to be our main concern."

5. Raimon Panikkar and the Unknown Christ

1. J. A. Lyons, *The Cosmic Christ in Origen and Teilhard de Chardin* (London: Oxford University Press, 1982), 49. Teilhard wrote: "In Scripture, Christ appears to me as essentially invested with the power to give the World, in him, *its definitive form.* He has been consecrated for a cosmic function." See also Henri de Lubac, *Teilhard de Chardin: The Man and His Meaning,* trans. René Hague (New York: Hawthorn Books, 1965), 31.

2. Lyons, *The Cosmic Christ,* 39.

3. For a discussion on evolution as a progression toward greater complexity and consciousness, see Timothy Jamison, "The Personalized Universe of Teilhard de Chardin," in *There Shall Be One Christ,* ed. Michael Meilach (New York: Franciscan Institute, 1968), 21-24.

4. Ursula King, *Towards a New Mysticism: Teilhard de Chardin & Eastern Religions* (New York: Seabury Press, 1980), 225.

5. King, *Towards a New Mysticism,* 225.

6. Although Ewert Cousins suggested this term to me in a private conversation, he points to a similar idea in his *Christ of the 21st Century* where he writes: "When Christian consciousness opens to global consciousness, a new type of systematic theology can be born. This new theology calls for a new kind of theologian with a new type of consciousness—a multidimensional, cross-cultural consciousness characteristic of the mutation of the Second Axial Period." See Cousins, *Christ of the 21st Century* (Rockport, Mass.: Element Books, 1992), 79.

7. Raimon Panikkar, *The Intra-Religious Dialogue* (New York: Paulist Press, 1978), 2. See also Gerard Hall, "Multi-Faith Dialogue in Conversation with Raimon Panikkar," Paper presented at the Australian Association for the Study of Religions, July 4-6, 2003, p. 2. See http://dlibrary. acu.edu.au/staffhome.

8. Ewert H. Cousins, "Raimundo Panikkar and the Christian Systematic Theology of the Future," *Cross Currents* 29 (1979): 143; Ewert

Cousins with Janet Kvamme Cousins, "Uniting Human, Cosmic and Divine: The Vision of Raimon Panikkar," *America* (January 1-8, 2007): 22.

9. Raimon Panikkar, *Christophany: The Fullness of Man*, trans. Alfred DiLascia (Maryknoll, N.Y.: Orbis Books, 2004), 4.

10. Panikkar, *Christophany*, 3-4.

11. Raimon Panikkar, "A Christophany for Our Time," *Theology Digest* 39, no. 1 (Spring 1992): 12.

12. Cited in Hall, "Multi-Faith Dialogue," 10.

13. Panikkar, "A Christophany for Our Time," 13 -14.

14. Panikkar, "A Christophany for Our Time," 3.

15. Cousins, "Uniting Human, Cosmic and Divine," 22.

16. Cousins, *Christ of the 21st Century*, 81.

17. Bonaventure, *Itinerarium Mentis in Deum* 6.5 (V, 311).

18. Panikkar, *Christophany*, 149; idem, "A Christophany for Our Time," 7-8.

19. Panikkar, "A Christophany for Our Time," 9.

20. Panikkar, "A Christophany for Our Time," 15.

21. Panikkar, "A Christophany for Our Time," 21.

22. Panikkar, "A Christophany for Our Time," 20.

23. For a discussion on Christ as the cosmotheandric mystery, see Panikkar, *Christophany*, 180-84; Cheriyan Menacherry, *Christ: The Mystery in History, A Critical Study on the Christology of Raymond Panikkar* (Frankfurt am Main: Peter Lang, 1996), 117-20; Paul F. Knitter, *No Other Name? A Critical Survey of Christian Attitudes Toward the World Religions* (Maryknoll, N.Y.: Orbis Books, 1985), 154-56.

24. Panikkar, "A Christophany for Our Time," 7; idem, *Christophany*, 147. Emphasis added. See also Raimon Panikkar, *The Unknown Christ of Hinduism* (Maryknoll, N.Y.: Orbis Books, 1981), 27, where he describes Christ as "a living symbol for the totality of reality: human divine, cosmic," or what he calls the "cosmotheandric reality."

25. Francis d'Sa, foreword to Panikkar, *Christophany*, xvi.

26. Panikkar, "A Christophany for Our Time," 11. Paul Knitter writes: "Panikkar clearly states that no historical name or form can be the full, final expression of the Christ. Christ, 'as the universal symbol for salvation cannot be objectified and thus reified as a merely historical personage.' ... Panikkar warns against an idolatrous form of historicism in Christianity," although he does recognize that Jesus is the ultimate form of Christ. See Knitter, *No Other Name?* 155-56.

27. Panikkar, "A Christophany for Our Time," 5.

28. Panikkar, "A Christophany for Our Time," 5.

29. Panikkar, *Christophany*, xx.

30. Panikkar, *Christophany*, 21.

31. Louis-Marie Chauvet, *Symbol and Sacrament: A Sacramental Reinterpretation of Christian Existence*, trans. Patrick Madigan and Madeleine Beaumont (Minneapolis: Liturgical Press, 1995), 112. See also Roger Haight, *Dynamics of Theology* (New York: Paulist Press, 1990), 132-42, who points out that there is an intrinsic connection between the symbol and that which is symbolized (p. 134).

32. Chauvet, *Symbol and Sacrament*, 112-13.

33. Chauvet, *Symbol and Sacrament*, 114-15.

34. Panikkar, *Christophany*, 144, 147.

35. Gustave Martelet, *The Risen Christ and the Eucharistic World*, trans. René Hague (New York: Seabury Press, 1976), 24.

36. Panikkar, *Christophany*, 162.

37. Panikkar, *Christophany*, 146.

38. Panikkar, *Christophany*, 146.

39. Panikkar, "A Christophany for Our Time," 4.

40. Panikkar, "A Christophany for Our Time," 4.

41. Teilhard de Chardin, *Phenomenon of Man*, trans. Bernard Wall (New York: Harper & Row, 1959), 262.

42. Panikkar, *Christophany*, 106.

43. Panikkar, *Christophany*, xvi.

44. Panikkar, *Christophany*, 101.

45. Cousins, "Uniting Human, Cosmic and Divine," 22.

46. These points are condensed from "A Christophany for Our Time," 6-20.

47. Panikkar, "A Christophany for Our Time," 6.

48. Panikkar, *Christophany*, 122.

49. Panikkar, *Christophany*, 124.

50. Panikkar, *Christophany*, 128.

51. Panikkar, "A Christophany for Our Time," 20.

52. Leonard Swidler, "The Age of Global Dialogue," in *Doors of Understanding: Conversations on Global Spirituality in Honor of Ewert Cousins*, ed. Steven L. Chase (Quincy, Ill.: Franciscan Press, 1997), 18.

53. Swidler, "Age of Global Dialogue," 20.

54. Peter Phan, "Praying to the Buddha: Living amid Religious Pluralism," *Commonweal* (January 26, 2007): 11.

55. Phan, "Praying to the Buddha," 11.

56. Raimon Panikkar, *Myth, Faith and Hermeneutics: Cross Cultural Studies* (New York: Paulist Press, 1980), 243.

57. Panikkar, *The Intra-Religious Dialogue*, 14.

58. Miroslav Volf, *Exclusion and Embrace: A Theological Exploration of Identity, Otherness, and Reconciliation* (Nashville: Abingdon Press, 1996), 141.

59. Volf, *Exclusion and Embrace*, 141.

60. Volf, *Exclusion and Embrace*, 142.

61. Volf, *Exclusion and Embrace*, 144.

62. For an explanation of embrace and not-understanding, see Volf, *Exclusion and Embrace*, 145-56.

63. For a discussion on poverty and prayer, see Ilia Delio, *Franciscan Prayer* (Cincinnati: St. Anthony Messenger Press, 2004), 77-90.

64. Panikkar, *Unknown Christ of Hinduism*, 35.

65. The idea of dialogical dialogue, according to Panikkar, differs from dialectical dialogue because it "stands on the assumption that nobody has access to the universal horizon of human experience, and that only by not postulating the rules of the encounter from a single side can one proceed towards a deeper and more universal understanding of oneself and thus come closer to one's own realization." See Panikkar, *Intra-Religious Dialogue*, 91.

66. Panikkar refers to this as a "human cosmic trust" or cosmotheandric confidence." See Raimon Panikkar, *Invisible Harmony* (Minneapolis: Fortress Press, 1995), 174-75.

67. Panikkar, *Unknown Christ of Hinduism*, 43. Panikkar writes: "A Christian will never fully understand Hinduism if he is not, in one way or another, converted to Hinduism. Nor will a Hindu ever fully understand Christianity unless he, in one way or another, becomes a Christian."

68. Panikkar, *Intra-Religious Dialogue*, 14.

69. Panikkar, *Myth, Faith and Hermeneutics*, 244.

70. Hall, "Multi-Faith Dialogue," 6.

71. Panikkar, *Unknown Christ of Hinduism*, 147-48.

72. Panikkar, *Intra-Religious Dialogue*, 70.

6. The Transcultural Christ: Thomas Merton and Bede Griffiths

1. Zachary Hayes, "Bonaventure of Bagnoregio: A Paradigm for Franciscan Theologians?" in *The Franciscan Intellectual Tradition*, ed. Elise Saggau (New York: Franciscan Institute, 2001), 56.

2. Phyllis Zagano, *Twentieth-Century Apostles: Contemporary Spirituality in Action* (Minneapolis: Liturgical Press, 1999), 91.

3. William M. Thompson, *Jesus, Lord and Savior: A Theopatic Christology and Soteriology* (New York: Paulist Press, 1980), 252.

4. Thomas Merton, *Conjectures of a Guilty Bystander* (New York: Doubleday Image, 1968), 21.

5. Thomas Merton, *New Seeds of Contemplation* (New York: New Directions, 1961), 34-36.

6. Thomas Merton, *No Man Is an Island* (New York: Harcourt Brace, 1955), 105.

7. Thomas Merton, *Introductions East and West* (Greensboro, N.C.: Unicorn Press, 1981), 111.

8. Thompson, *Jesus, Lord and Savior*, 254.

9. Thompson, *Jesus, Lord and Savior*, 256.

10. Bonaventure's journey begins on the level of creation, moves into the inner world of the soul and then to the contemplation of God.

11. Thompson, *Jesus, Lord and Savior*, 259.

12. Merton, *New Seeds of Contemplation*, 53.

13. Merton, *Introductions East and West*, 92.

14. Thomas Merton, "Notes for a Philosophy of Solitude," in *Disputed Questions* (New York: Farrar, Straus & Cudahy, 1960), 207.

15. Merton, *Disputed Questions*, 168.

16. Thomas Merton, *Faith and Violence* (Notre Dame, Ind.: Notre Dame Press, 1968), 224.

17. Merton, *New Seeds of Contemplation*, 53.

18. Merton, *New Seeds of Contemplation*, 70-79.

19. Thomas Merton, *Contemplation in a World of Action* (New York: Image Books, 1973), 225-226.

20. Thompson, *Jesus, Lord and Savior*, 260.

21. For a discussion of the Christ mysticism of the *Canticle*, see Ilia Delio, "The Canticle of Brother Sun: A Song of Christ Mysticism," *Franciscan Studies* 52 (1992): 1-22; idem, *A Franciscan View of Creation: Learning to Live in a Sacramental World*, vol. 2, *The Franciscan Heritage Series*, ed. Joseph P. Chinnici (New York: Franciscan Institute, 2003), 17-20.

22. Ken Wilber, *Up from Eden: A Transpersonal View of Human Evolution* (Wheaton, Ill.: Quest Books, 1996), 15.

23. Wilber, *Up from Eden*, 13.

24. William M. Thompson, "The Risen Christ, Transcultural Consciousness, and the Encounter of the World Religions," *Theological Studies* 37 (1976): 401.

25. Merton, *Contemplation in a World of Action*, 229-30.

26. Thompson, "Risen Christ," 401.

27. Merton, "The Power and Meaning of Love," in *Disputed Questions*, 123-25.

28. William Johnston, *Silent Music: The Science of Meditation* (New York: Harper & Row, 1974), 160.

29. For a discussion on Christ as coincidence of opposites, see Ewert H. Cousins, *Bonaventure and the Coincidence of Opposites* (Chicago: Franciscan Herald Press, 1978); idem, "The Coincidence of Opposites in the Christology of Saint Bonaventure," *Franciscan Studies* 28 (1968): 27-45.

30. For an introduction to the life and thought of Bede Griffiths, see Wayne Teasdale, "Introduction: Bede Griffiths as Visionary Guide," in *The*

Other Half of My Soul: Bede Griffiths and the Hindu-Christian Dialogue, compiled by Beatrice Bruteau (Wheaton, Ill.: Quest Books, 1996), 2-24. There are several biographies of Bede Griffiths, including Kathryn Spink's *A Sense of the Sacred: A Biography of Bede Griffiths* (Maryknoll, N.Y.: Orbis Books, 1989), and Shirley du Boulay, *Beyond the Darkness: A Biography of Bede Griffiths* (New York: Doubleday, 1998).

31. Bede Griffiths, *The Marriage of East and West* (Springfield, Ill.: Templegate, 1982), 8.

32. Griffiths, *Marriage of East and West*, 152.

33. Bede Griffiths, *The Cosmic Revelation: The Hindu Way to God* (Springfield, Ill.: Templegate, 1983), 112.

34. Bede Griffiths, "The One Mystery," *The Tablet* (March 9, 1974): 223.

35. Bede Griffiths, *A New Vision of Reality: Western Science, Eastern Mysticism and Christian Faith* (Springfield, Ill.: Templegate, 1990), 264ff.

36. Wayne Teasdale, "Christianity and the Eastern Religions: The Possibility of Mutual Growth," in *The Other Half of My Soul*, 156-57. Teasdale follows closely Bede's thoughts on interreligious dialogue. In a talk on interreligious dialogue Bede said, "What we find is that if you're arguing doctrines and so on, you get nowhere, but when you meet in meditation you begin to share your own inner experience [and] you begin to realize an underlying unity behind the religions" (Teasdale, "Christianity and the Eastern Religions," 131).

37. Frank Gerry, "Dom Bede Griffiths—Mystic of the East: A Human Search," *Australian EJournal of Theology* 8 (October 2006). http://dlibrary.acu.edu.au/research/theology/ejournal.

38. Bede Griffiths, *Return to the Center* (Springfield, Ill.: Templegate, 1976), 36, 71.

39. Griffiths, *Return to the Center*, 71.

40. Griffiths, *Return to the Center*, 73.

41. Gerry, "Dom Bede Griffiths—Mystic of the East."

42. Griffiths, *Return to the Center*, 73-74.

43. Gerry, "Dom Bede Griffiths—Mystic of the East."

44. Raimon Pannikar, "Hinduism & Christ," in *In Spirit and in Truth: Essays Dedicated to Father Ignatius Hirudayan, S.J.* (Madras: Aikaya Alayam, 1985), 115.

45. Bede Griffiths, "The New Consciousness," *The Tablet* (January 16, 1993): 70.

46. Gerry, "Dom Bede Griffiths—Mystic of the East."

47. Griffiths, *Return to the Center*, 144.

48. Griffiths, *Return to the Center*, 145.

49. Griffiths, *Return to the Center*, 31.

50. Griffiths, *Cosmic Revelation*, 127.

51. Griffiths, *Return to the Center*, 32.

52. Griffiths, *Return to the Center*, 31.

53. Griffiths, *Return to the Center*, 32.

54. Griffiths, *Return to the Center*, 40.

55. Griffiths, *Return to the Center*, 60.

56. Griffiths, *Return to the Center*, 51.

57. Griffiths, *Return to the Center*, 52.

58. Griffiths, *Return to the Center*, 52.

59. Griffiths, *A New Vision of Reality*, 264.

60. Griffiths, *Return to the Center*, 61.

61. Gerry, "Dom Bede Griffiths—Mystic of the East."

7. Christology Reborn

1. Pierre Teilhard de Chardin, *Christianity and Evolution*, trans. René Hague (New York: Harcourt Brace Jovanovich, 1971), 94-95.

2. Bernard McGinn first introduced the concept of vernacular theology in *Meister Eckhart and the Beguine Mystics* (New York: Continuum, 1983), 4-14, and has further developed it in his *Flowering of Mysticism*, vol. 3, *The Presence of God: A History of Western Mysticism* (New York: Crossroad, 1998), 18-24.

3. Paul F. Knitter, *No Other Name? A Critical Survey of Christian Attitudes Toward the World Religions* (Maryknoll, N.Y.: Orbis Books, 1985), 187.

4. The term "perichoresis" was first used by the eighth-century theologian John Damascene, who said that the divine persons of the Trinity are not only related to one another but mutually inhere in one another and draw life from one another. Bonaventure was influenced by the idea of perichoresis but used the Latin instead, *circumincessio*, which means that the divine persons "move around one another" in a communion of love. See Ilia Delio, *Simply Bonaventure: An Introduction to His Life, Thought and Writings* (New York: New City Press, 2001), 41.

5. Karl Rahner, "The Two Basic Types of Christology," *Theological Investigations*, vol. 13, trans. David Bourke (New York: Seabury Press, 1975), 215-16.

6. Raimon Panikkar, *Christophany: The Fullness of Man*, trans. Alfred DiLascia (Maryknoll, N.Y.: Orbis Books, 2004), 120.

7. Panikkar, *Christophany*, 144, 147.

8. Panikkar, *Christophany*, 150. Panikkar writes, "Jesus Christ as undivided experience constitutes the central Christian dogma. The copula 'is' collapses: otherwise, it would introduce an epistemic split of the unity of that experience … Jesus is Christ, but Christ cannot be identified com-

pletely with Jesus of Nazareth." In the foreword to this book, Francis d'Sa writes: "The Christ of the christophany is, for example, the Christ that was, is, and will be at work in the whole of creation, that is, in every single being and not only in Jesus. Jesus is Christ but Christ cannot be identified completely with Jesus" (p. xvi).

9. Panikkar, *Christophany*, 147.

10. Karl Rahner, "Current Problems in Christology," *Theological Investigations*, vol. 1, trans. Cornelius Ernst (Baltimore: Helicon Press, 1961), 192.

11. See Ewert Cousins, "The Coincidence of Opposites in the Christology of Saint Bonaventure," *Franciscan Studies* 28 (1968): 27-34; idem, *Bonaventure and the Coincidence of Opposites* (Chicago: Franciscan Herald Press, 1978).

12. Bonaventure, *Itin.* 6.5 (V, 311). Engl. trans. Cousins, *Bonaventure and the Coincidence of Opposites*, 107.

13. For a discussion on the concept of "gazing," see Ilia Delio, *Franciscan Prayer* (Cincinnati: St. Anthony Messenger Press, 2004), 77-90.

14. Bonaventure, *Itin.* 4.8 (V, 308).

15. Cited in William M. Thompson, *Jesus, Lord and Savior: A Theopatic Christology and Soteriology* (New York: Paulist Press, 1980), 252.

16. Cited in Thompson, *Jesus, Lord and Savior*, 256.

17. Thomas Merton, *Contemplation in a World of Action* (New York: Image Books, 1973), 225.

18. George Maloney, *The Cosmic Christ: From Paul to Teilhard* (New York: Sheed & Ward, 1968), 184.

19. Maloney, *Cosmic Christ*, 184-85.

20. There is no theology that is "neutral" or completely objective apart from some type of subjective experience or view of the world. As the Dominican scholar Marie-Dominique Chenu wrote, "The fact is in the final analysis theological systems are simply the expressions of a spirituality. It is this that gives them their interest and their grandeur ... One does not get to the heart of a system via the logical coherence of its structure or the plausibility of its conclusions. One gets to that heart by grasping it in its origins via that fundamental intuition that serves to guide a spiritual life and provides the intellectual regimen proper to that life." Cited in Gustavo Gutiérrez, *We Drink from Our Own Wells*, trans. Matthew O' Connell (Maryknoll, N.Y.: Orbis Books, 1984), 147 n. 2. In light of Chenu's insight, theology in the second axial period must be born out of a deep inner center that incorporates the lived experience and dynamism of the world.

21. Emmanuel Falque, "The Phenomenological Act of *Perscrutatio* in the Proemium of St. Bonaventure's Commentary on the Sentences," trans. Elisa Mangina, *Medieval Philosophy and Theology* 10 (2001): 9.

22. Falque, "Phenomenological Act of *Perscrutatio*," 6.

23. Falque, "Phenomenological Act of *Perscrutatio*," 7.

24. Jean Luc Marion, *Reduction et donation: Recherches sur Husserl, Heidegger, et la phénoménologie* (Paris: Presses Universitaires de France, 1989), 63.

25. Falque, "Phenomenological Act of *Perscrutatio*," 18.

26. Bonaventure, *Itin.* 7.6 (V, 313). Engl. trans. Cousins, *Bonaventure and the Coincidence of Opposites*, 115.

27. Bonaventure, *Hex.* 2.32 (V, 342). Engl. trans., José de Vinck, *Collations on the Six Days*, vol. 5, *The Works of Bonaventure* (Paterson, N.J.: St. Anthony Guild Press, 1970).

28. Bonaventure, *Hex.* 2.32 (V, 342). Engl. trans. de Vinck, *Collations on the Six Days*, 39.

29. There have been very few studies on the role of the Holy Spirit in Bonaventure's theology, especially the role of the Spirit in his cosmic Christology. A particularly rich area for this study is Bonaventure's eschatology, especially as it is laid out in his collations on the six days of creation (*Collationes in Hexaëmeron*). A seminal work on Bonaventure's eschatology is Joseph Ratzinger's *Theology of History in St. Bonaventure*, trans. Zachary Hayes (Chicago: Franciscan Press, 1971). For a study of the Holy Spirit in Bonaventure's early writings, see Zachary Hayes, "The Doctrine of the Spirit in the Early Writings of St. Bonaventure," in *Doors of Understanding: Conversations on Global Spirituality in Honor of Ewert Cousins*, ed. Steven L. Chase (Quincy, Ill.: Franciscan Press, 1997), 179-98.

30. See Henri de Lubac, *Teilhard de Chardin: The Man and His Meaning*, trans. René Hague (New York: Hawthorn Books, 1965), 188.

8. Co-creators in Christ

1. George Maloney, *The Cosmic Christ: From Paul to Teilhard* (New York: Sheed and Ward, 1968), 189.

2. See Philip Hefner, *The Human Factor: Evolution, Culture and Religion* (Minneapolis: Fortress Press, 1993), 23-51, 255-75. See also Philip Hefner, "The Evolution of the Created Co-Creator," in *An Evolving Dialogue: Theological and Scientific Perspectives on Evolution*, ed. James B. Miller (Harrisburg, Pa.: Trinity Press International, 2001), 399-416.

3. Zachary Hayes, "The Meaning of *Convenientia* in the Metaphysics of St. Bonaventure," *Franciscan Studies* 34 (1974): 83.

4. Maloney, *Cosmic Christ*, 211.

5. Maloney, *Cosmic Christ*, 211.

6. Pierre Teilhard de Chardin, *The Divine Milieu: An Essay on the Interior Life*, trans. William Collins (New York: Harper & Row, 1960), 66.

7. This is the thesis of Teilhard's classic *Divine Milieu*. See also his "My Universe" in *Process Theology*, ed. Ewert H. Cousins (New York: Newman Press, 1971), 249-55.

8. Ursula King, *Christ in All Things: Exploring Spirituality with Teilhard de Chardin* (Maryknoll, N.Y.: Orbis Books, 1997), 93.

9. King, *Christ in All Things*, 93.

10. Pierre Teilhard de Chardin, *Christianity and Evolution*, trans. René Hague (New York: Harcourt Brace Jovanovich, 1971), 93.

11. Teilhard de Chardin, *Christianity and Evolution*, 91-92.

12. Teilhard de Chardin, *Christianity and Evolution*, 95.

13. Henri de Lubac, *Teilhard de Chardin: The Man and His Meaning*, trans. René Hague (New York: Hawthorn Books, 1965), 56-61. See also Thomas King, *Teilhard's Mass: Approaches to "The Mass on the World"* (New York: Paulist Press, 2005).

14. Maloney, *Cosmic Christ*, 211.

15. Bede Griffiths, *Return to the Center* (Springfield, Ill.: Templegate, 1976), 116.

16. Teilhard de Chardin, *Christianity and Evolution*, 85.

17. De Lubac, *Teilhard de Chardin*, 101 n. 9. In chapter 13 of his book, de Lubac discusses the limitations of Teilhard's work. The problem of evil in light of an optimistic view of evolution and Christ in evolution requires a much more comprehensive examination than what I can provide here. Teilhard encountered resistance to his doctrine precisely because his critics maintained that he did not sufficiently account for evil in the process of Christogenesis; however, such criticism fails to grasp the depth of Teilhard's thought. For an interesting study of evil in Teilhard's doctrine, see Thomas Mulvihill King, *Teilhard, Evil and Providence* (Chambersburg, Pa.: Anima Publications, 1989); Robert Faricy, *The Spirituality of Teilhard de Chardin* (Minneapolis: Winston Press, 1981), 47-55.

18. Teilhard de Chardin, *The Phenomenon of Man*, trans. Bernard Wall (New York: Harper & Row, 1959), 265.

19. Teilhard de Chardin, *Christianity and Evolution*, 184.

20. Teilhard de Chardin, *Christianity and Evolution*, 184-85.

21. Thomas Merton, "The Power and Meaning of Love," in *Disputed Questions* (New York: Farrar, Straus & Cudahy, 1960), 123-25.

22. Fritjof Capra, *The Tao of Physics* (Berkeley, Calif.: Shambala, 1975), 142.

23. David Bohm, *Wholeness and the Implicate Order* (New York: Routledge & Kegan Paul, 1980), 124-25; William Stoeger, "What Does Science Say About Creation?" *The Month* (August/September 1988): 807.

24. Beatrice Bruteau, *The Grand Option: Personal Transformation and a New Creation* (Notre Dame, Ind.: University of Notre Dame Press, 2001), 172-73.

25. Thomas Merton, quoted in James Forest, *Living with Wisdom: A Life of Thomas Merton* (Maryknoll, N.Y.: Orbis Books, 1991), 216.

26. Griffiths, *Return to the Center*, 60. Actually Bede borrows this insight from the Muslims.

27. Thomas Merton, *New Seeds of Contemplation* (New York: New Directions, 1961), 75.

28. Bruteau, *Grand Option*, 171.

29. For a good discussion of the power of the word among the desert fathers, see Douglas Burton-Christie, *The Word in the Desert: Scripture and the Quest for Holiness in Early Christian Monasticism* (New York: Oxford University Press, 1993), 107-29.

30. Walter Ong, *Orality and Literacy: The Technologizing of the Word* (London: Methuen, 1982), 32-33.

31. John Borelli, "A Necessary Dialogue," *America* (February 5, 2007): 12-13.

32. Vatican translation of the message Benedict XVI sent September 4 to Archbishop Domenico Sorrentino of Assisi-Nocera, on the occasion of the twentieth anniversary of the Interreligious Meeting of Prayer for Peace (October 9, 2006) (www.zenit.org/english).

33. Raimon Panikkar, *The Intra-Religious Dialogue* (New York: Paulist Press, 1978), xxiif.

34. Gerard Hall, "Multi-Faith Dialogue in Conversation with Raimon Panikkar," Paper presented at the Australian Association for the Study of Religions, July 4-6, 2003, p. 2. See http://dlibrary.acu.edu.au/staffhome, 6.

35. Panikkar refers to this as "human cosmic trust" or "cosmotheandric confidence." See his *Invisible Harmony* (Minneapolis: Fortress Press, 1995), 174ff.

36. Hall, "Multi-Faith Dialogue," 5; Raimon Panikkar, *The Unknown Christ of Hinduism* (Maryknoll, N.Y.: Orbis Books, 1981), 35.

37. Panikkar, *Intra-Religious Dialogue*, 61f.

38. Hall, "Multi-Faith Dialogue," 6-7.

39. William M. Thompson, *Jesus, Lord and Savior: A Theopatic Christology and Soteriology* (New York: Paulist Press, 1980), 268.

40. King, *Christ in All Things*, 80.

41. Zachary Hayes, "Christ, Word of God and Exemplar of Humanity," *Cord* 46.1 (1996): 12.

42. Hayes, "Christ, Word of God and Exemplar of Humanity," 15.

43. Zachary Hayes, "Christology-Cosmology," in *Spirit and Life: A Journal of Contemporary Franciscanism*, vol. 7, ed. Anthony Carrozzo, Kenneth Himes, and Vincent Cushing (New York: Franciscan Institute, 1997), 52.

44. Cited in Margaret Wheatley, *Leadership and the New Science: Learn-*

ing about Organization from an Orderly Universe (San Francisco: Berrett-Koehler Publishers, 1992), 74.

45. Ken Wilber, *Up from Eden: A Transpersonal View of Human Evolution* (Wheaton, Ill.: Quest Books, 1996), 16.

46. King, *Christ in All Things*, 158.

47. For Teilhard's defense of his writings, see de Lubac, *Teilhard de Chardin*, 186-203. A monitum on Teilhard's theological writings was issued by the Holy See on June 30, 1962, and reiterated on July 20, 1981.

48. Karl Rahner, "Christian Living Formerly and Today" in *Further Theology of the Spiritual Life* I, *Theological Investigations,* vol. 7, trans. David Bourke (New York: Herder & Herder, 1971), 14f. See Gregory LaNave, *Through Holiness to Wisdom: The Nature of Theology according to St. Bonaventure* (Rome: Instituto Storico Dei Cappuccini, 2005), 8, who writes, "In this essay, Rahner warned that a mere 'rational attitude to the speculative problems of the divine' was not enough, that 'the devout Christian of the future will either be a mystic, one who has experienced something, or he will cease to be anything at all.'"

9. Christ in Evolution: Technology and Extraterrestrial Life

1. Pierre Teilhard de Chardin, *Phenomenon of Man*, trans. Bernard Wall (New York: Harper & Row, 1959), 262.

2. Teilhard de Chardin, *Phenomenon of Man*, 290.

3. Karl Rahner, *On the Theology of Death*, trans. Charles H. Henkey (New York: Herder & Herder, 1961), 65; Denis Edwards, *Jesus and the Cosmos* (New York: Paulist Press, 1991), 100.

4. Karl Rahner, *The Theology of the Spiritual Life*, trans. Karl-J. and Boniface Kruger, vol. 3, *Theological Investigations* (Baltimore: Helicon Press, 1967), 44.

5. Summarizing Rahner's thoughts, Edwards states that "what we do in our history has final and definitive value." See Edwards, *Jesus and the Cosmos*, 97.

6. Bonaventure, *Hex.* 20.15; 22.23 (V, 428, 441). Bonaventure writes, [In the final age] "there will be a consummation of those sufferings of Christ which he now bears in his [mystical] body." And further on he writes, "Christ [must] appear and suffer in his mystical body." Engl. trans., José de Vinck, *Collations on the Six Days*, vol. 5, *The Works of Bonaventure* (Paterson, N.J.: St. Anthony Guild Press, 1970), 309, 352.

7. George Maloney, *The Cosmic Christ: From Paul to Teilhard* (New York: Sheed & Ward, 1968), 217.

8. Jean Danielou, "The Meaning and Significance of Teilhard de Chardin," trans. John Lyon, *Communio* 15 (Fall 1988): 355.

9. Fritjof Capra, *The Web of Life: A New Scientific Understanding of Living Systems* (New York: Anchor Books, 1997), 36-50; Rupert Sheldrake, "Mysticism and the New Science," in *The Other Half of My Soul: Bede Griffiths and the Hindu-Christian Dialogue*, compiled by Beatrice Bruteau (Wheaton, Ill.: Quest Books, 1996), 336-355.

10. On the future of the universe, see Mark William Worthing, *God, Creation and Contemporary Physics* (Minneapolis: Fortress Press, 1996), 159-75; Freeman Dyson, *Infinite in All Directions* (New York: Harper & Row, 1988).

11. Dyson, *Infinite in All Directions*, 107. Dyson writes, "Life resides in organization rather than substance. I am assuming that my consciousness is inherent in the way the molecules in my head are organized, not in the substance of the molecules themselves. If this assumption is true, that life is organization rather than substance, then it makes sense to imagine life detached from flesh and blood and embodied in networks of superconducting circuitry or in interstellar dust clouds."

12. David F. Noble, *The Religion of Technology: The Divinity of Man and the Spirit of Invention* (New York: A. A. Knopf, 1997), 14.

13. Maloney, *Cosmic Christ*, 218-19.

14. Noble, *Religion of Technology*, 5.

15. Noble, *Religion of Technology*, 143-71. For example A.I. guru Earl Cox has argued that exponentially accelerating advances in science and technology have sped up the course of evolution, outdistancing their creators. We are at the dawn of a new robotic "supercivilization" which will remake the entire universe in its digital image. "Technology will soon enable human beings to change into something else altogether," he states, and thereby "escape the human condition ... We will download our minds into vessels created by our machine children and with them, explore the universe" (p. 164).

16. Ray Kurzweil, *The Singularity Is Near: When Humans Transcend Biology* (New York: Penguin Books, 2006).

17. See, e.g., Ray Kurzweil, *The Age of Spiritual Machines: When Computers Exceed Human Intelligence* (New York: Penguin Books, 1999), 61-156. The interested futurist may also want to look at Kurzweil's website, KurzweilAI.net.

18. "The Singularity: A Talk with Ray Kurzweil." www.edge.org.

19. See, e.g., Michael Heim's claim that "our fascination with computers ... is more deeply spiritual than utilitarian." "When on-line," he writes, "we break free from bodily existence ... what better way to emulate God's knowledge than to generate a virtual world constituted by bits

of information. Over such a cyber world human beings could enjoy a god-like instant access." Cited in Noble, *Religion of Technology*, 159.

20. Noble, *Religion of Technology*, 162. Hans Moravec is a researcher in robotics at Carnegie Mellon University. His books include *Mind Children: The Future of Robot and Human Intelligence* (Cambridge: Harvard University Press, 1990), and *Robot: Mere Machine to Transcendent Mind* (New York: Oxford University Press, 2000).

21. Noreen Herzfeld, *In Our Image: Artificial Intelligence and the Human Spirit* (Minneapolis: Augsburg Fortress Press, 2002), 304.

22. Noreen Herzfeld, "Creating in Our Own Image: Artificial Intelligence and the Image of God," *Zygon* 37, no. 2 (June 2002): 304.

23. Philip Hefner, "Technology and Human Becoming," *Zygon* 37, no. 3 (September 2002): 658.

24. For a theological discussion of *techno sapiens,* see Antje Jackelén, "The Image of God as *Techno Sapiens*," *Zygon* 37, no. 2 (June 2002): 289-301.

25. Ilia Delio, "Artificial Intelligence and Christian Salvation: Compatibility or Competition?" *New Theology Review* 16, no. 4 (November 2003): 39-51.

26. Philip Hefner, *Technology and Human Becoming* (Minneapolis: Fortress Press, 2003), 75. Hefner notes that "technosapien" is a more recent coinage that aims at the same idea as "cyborg." Some futurists now speak of "cyber sapiens," saying, "We will no longer be Homo sapiens, but Cyber sapiens—a creature part digital and part biological that will have placed more distance between its DNA and the destinies they force upon us than any other animal ... a creature capable of steering our own evolution." See KurzweilAI.net.

27. Naomi Goldenberg, *Resurrecting the Body: Feminism, Religion and Psychoanalysis [Returning Words to Flesh]* (New York: Crossroad, 1993), 11.

28. Hefner, *Technology and Human Becoming,* 77.

29. Hefner, *Technology and Human Becoming,* 78.

30. Hefner, *Technology and Human Becoming,* 84-85.

31. Hefner, *Technology and Human Becoming,* 86-87. Hefner raises an interesting question: "What is the significance of the fact that on our planet, at least, God has set up a system in which the creatures who transcend humans in the chain of evolution may be creatures we have designed and created, so that their act of transcending us is at the same time our own act of transcending ourselves?" (p. 80).

32. J. A. Lyons, *The Cosmic Christ in Origen and Teilhard de Chardin* (London: Oxford University Press, 1982), 41.

33. Lyons, *Cosmic Christ in Origen and Teilhard de Chardin,* 41.

34. Andrew J. Burgess, "Earth Chauvinism," *Christian Century* 93 (1976): 1098.

35. Burgess, "Earth Chauvinism," 1098.

36. Burgess, "Earth Chauvinism," 1100.

37. Christen Brownlee, "Is Anybody Out There?" *Science News* 169, no. 3 (2006), p. 42. Online at http://www.sciencenews.org/articles/20060121/bob10.asp.

38. Lewis S. Ford, *The Lure of God: A Biblical Background for Process Theism* (Philadelphia: Fortress Press, 1978), 56.

39. J. Edgar Burns, "Cosmolatry," *The Catholic World* 191 (1960): 284.

40. Burgess, "Earth Chauvinism," 1098.

41. Ted Peters, *Science, Theology, and Ethics* (Burlington, Vt.: Ashgate, 2003), 128.

42. For a discussion of the cosmic Christology of the Greek fathers, see Maloney, *Cosmic Christ*, 142-81.

43. Zachary Hayes, "Christ, Word of God and Exemplar of Humanity," *Cord* 46, no. 1 (1996): 3.

44. The term "christophic principle" is a play on the "anthropic principle," which suggests that the universe is fine-tuned for human life (strong version) or at least has the necessary ingredients for life, even if by chance (weak version). See John Leslie, *Universes* (New York: Routledge, 1996); John D. Barrow and Frank J. Tipler, *The Anthropic Cosmological Principle* (New York: Oxford University Press, 1988). I am indebted to Robert Ulanowicz, who coined the expression "christophic principle."

45. Bonaventure, *Sermo II in nat. Dom.* (IX, 107); trans. "Sermon II on the Nativity," in *What Manner of Man? Sermons on Christ by St. Bonaventure*, 2nd ed., trans. Zachary Hayes (Chicago: Franciscan Herald Press, 1989), 74.

46. Hayes, "Christ, Word of God and Exemplar of Humanity," 48.

47. Ilia Delio, "Revisiting the Franciscan Doctrine of Christ," *Theological Studies* 64 (2003): 14-15.

48. See, e.g., Daryl Dooming, "Evolution, Evil and Original Sin," *America* 185, no. 15 (November 12, 2001): 14-21; idem, *Doing without Adam and Eve: Sociobiology and Original Sin* (Minneapolis: Augsburg Fortress Press, 2001); Philip Hefner, *The Human Factor: Evolution, Culture and Religion* (Minneapolis: Fortress Press, 1993), 125-40.

49. The theological argument here is that wherever there is a created order there will be a capacity to love within that order since God, who is creator, is love. Thus, any creation that emerges out of trinitarian love will bear the pattern of that love according to its created order; hence, the pattern of love, following the model of the Trinity, must be relational. Thus, we speculate that even a species with superior intelligence will be able to love; such love will be communicated or expressed and will be the basis of its salvation.

50. Karl Rahner, "Natural Science and Reasonable Faith," *Theological*

Investigations, vol. 12, trans. Hugh M. Riley (New York: Crossroad, 1988), 51. Rahner uses the word "Logos" to describe the creative principle of God, but this principle is the second divine person of the Trinity and hence Word.

51. Paul Tillich, *Systematic Theology,* 3 volumes (Chicago: University of Chicago Press, 1951-63), 3:95f.

52. Raimon Panikkar, *Christophany: The Fullness of Man,* trans. Alfred DiLascia (Maryknoll, N.Y.: Orbis Books, 2004), 150. In the foreword to *Christophany* Francis d'Sa writes, "The Christ of the christophany is, for example, the Christ that was, is, and will be at work in the whole of creation, that is, in every single being and not only in Jesus. Jesus is Christ but Christ cannot be identified completely with Jesus" (p. xvi).

53. Panikkar, *Christophany,* 147.

54. John L. McKenzie, *Dictionary of the Bible* (New York: Macmillan, 1965), 436.

55. Karl Rahner, "The Two Basic Types of Christology," *Theological Investigations,* vol. 13, trans. David Bourke (New York: Seabury Press, 1975), 215-16.

56. Panikkar, *Christophany,* 120.

57. In terrestrial creation, Christ unites and transforms the created order in God through his death and resurrection. Bonaventure views the cross of Christ as the fullest expression of the triune God of love in history. That which God is—a communion of love—culminates in the complete self-offering of Jesus on the cross by which we are reconciled to God and completed in divine love. Would the cross be relevant to extraterrestrial life as well? It is difficult to say. Our logic of thought leads us to suggest that it is not the cross in itself that restores and completes creation in God; rather, it is that which most completely expresses the love of God in a personal act of love by which creatures/creation may be fully reconciled and completed in God. In terrestrial life, the cross is the perfect symbol of divine love and reconciliation; however, in another created order, a more relevant symbol may express divine love in a way that is appropriate to that order.

Conclusion

1. Bruno Barnhart, "The Revolution of Jesus," at www.bedegriffiths.com.

2. Pierre Teilhard de Chardin, *Christianity and Evolution,* trans. René Hague (New York: Harcourt Brace Jovanovich, 1971), 89. Emphasis added.

3. *A Human Search: Bede Griffiths Reflects on His Life*, ed. John Swindells, foreword by Wayne Teasdale (Ligouri, Mo.: Triumph Books, 1997), 108.

4. Barnhart, "The Revolution of Jesus," www.bedegriffiths.com.

5. Barnhart, "The Revolution of Jesus," www.bedegriffiths.com.

6. Luiz Carlos Susin, "Introduction: Emergence and Urgency of the New Pluralist Paradigm," trans. Paul Burns, in *Pluralist Theology: The Emerging Paradigm*, ed. Luiz Carlos Susin, Andrés Torres Queiruga, José María Vigil, *Concilium*, vol. 1 (London: SCM Press, 2007), 8.

7. Barnhart, "The Revolution of Jesus," www.bedegriffiths.com.

8. Francis d'Sa, foreword to Raimon Panikkar, *Christophany: The Fullness of Man*, trans. Alfred DiLascia (Maryknoll, N.Y.: Orbis Books, 2004), xx.

9. Denis Edwards, *Jesus and the Cosmos* (New York: Paulist Press, 1991), 97.

10. Teilhard de Chardin, *Christianity and Evolution*, 175.

SELECT BIBLIOGRAPHY

Barbour, Ian. *Issues in Science and Religion.* London: SCM Press, 1997.

Bohm, David. *Wholeness and the Implicate Order.* New York: Routledge & Kegan Paul, 1980.

Bonnefoy, Jean-François. *Christ and the Cosmos.* Translated by Michael D. Meilach. Paterson, N.J.: St. Anthony Guild, 1965.

Borg, Marcus J., and N. T. Wright. *The Meaning of Jesus: Two Visions.* New York: HarperCollins, 1999.

Bruteau, Beatrice. *The Grand Option: Personal Transformation and a New Creation.* Notre Dame, Ind.: University of Notre Dame Press, 2001.

———, ed. *The Other Half of My Soul: Bede Griffiths and the Hindu-Christian Dialogue.* Wheaton, Ill.: Quest Books, 1996.

Capra, Fritjof. *The Tao of Physics.* Berkeley, Calif.: Shambala, 1975.

———. *The Web of Life: A New Scientific Understanding of Living Systems.* New York: Anchor Books, 1997.

Chase, Steven L., ed. *Doors of Understanding: Conversations on Global Spirituality in Honor of Ewert Cousins.* Quincy, Ill.: Franciscan Press, 1997.

Cobb, John. *The Structure of Christian Existence.* New York: Seabury Press, 1979.

Cousins, Ewert H. *Bonaventure and the Coincidence of Opposites.* Chicago: Franciscan Herald Press, 1978.

———. *Christ of the 21st Century.* Rockport, Mass.: Element Books, 1992.

———, ed. *Process Reality.* New York: Newman Press, 1971.

de Lubac, Henri. *Teilhard de Chardin: The Man and His Meaning.* Translated by René Hague. New York: Hawthorn Books, 1965.

du Boulay, Shirley. *Beyond the Darkness: A Biography of Bede Griffiths.* New York: Doubleday, 1998.

Dunn, James D. G. *Christology in the Making.* Philadelphia: Westminster, 1980.

———. *The Theology of Paul the Apostle.* Grand Rapids: Eerdmans, 1997.

Dyson, Freeman. *Infinite in All Directions.* New York: Harper & Row, 1988.

Edwards, Denis. *Jesus and the Cosmos.* New York: Paulist Press, 1991.

Faricy, Robert. *The Spirituality of Teilhard de Chardin.* Minneapolis: Winston Press, 1981.

Fitzmyer, Joseph. *Christological Catechism : New Testament Answers*. New York: Paulist Press, 1991.

———. *The One Who Is to Come*. Grand Rapids: Eerdmans, 2006.

Gray, Donald P. *The One and the Many: Teilhard de Chardin's Vision of Unity*. New York: Herder & Herder, 1969.

Griffiths, Bede. *The Cosmic Revelation: The Hindu Way to God*. Springfield, Ill.: Templegate, 1983.

———. *The Marriage of East and West*. Springfield, Ill.: Templegate, 1982.

———. *A New Vision of Reality: Western Science, Eastern Mysticism and Christian Faith*. Springfield, Ill.: Templegate, 1990.

———. *Return to the Center*. Springfield, Ill.: Templegate, 1976.

Grillmeier, Aloys. *Christ in the Christian Tradition*. Vol. 1. *From the Apostolic Age to Chalcedon (451)*. Translated by John Bowden. Second revised edition. Atlanta: John Knox Press, 1975.

Haight, Roger. *Jesus Symbol of God*. Maryknoll, N.Y.: Orbis Books, 1999.

Haught, John F. *Christianity and Science: Towards a Theology of Nature*. Maryknoll, N.Y.: Orbis Books, 2007.

———. *God beyond Darwin: A Theology of Evolution*. Boulder, Colo.: Westview Press, 2000.

———. *Responses to 101 Questions on God and Evolution*. New York: Paulist Press, 2001.

Hawking, Stephen W. *A Brief History of Time*. New York: Bantam Books, 1998.

Hayes, Zachary. *The Hidden Center: Spirituality and Speculative Christology in St. Bonaventure*. New York: Franciscan Institute, 1992.

———. *A Window to the Divine: A Study of Christian Creation Theology*. Quincy, Ill.: Franciscan Press, 1997.

Hefner, Philip. *The Human Factor: Evolution, Culture and Religion*. Minneapolis: Fortress Press, 1993.

———. *Technology and Human Becoming*. Minneapolis: Fortress Press, 2003.

Herzfeld, Noreen. *In Our Image: Artificial Intelligence and the Human Spirit*. Minneapolis: Augsburg Fortress Press, 2002.

King, Thomas M. *Teilhard's Mass : Approaches to "The Mass on the World."* New York: Paulist Press, 2005.

———. *Teilhard's Mysticism of Knowing*. New York: Seabury Press, 1981.

King, Ursula. *Christ in All Things*. Maryknoll, N.Y.: Orbis Books, 1997.

———. *Towards a New Mysticism: Teilhard de Chardin & Eastern Religions*. New York: Seabury Press, 1980.

Knitter, Paul F. *No Other Name? A Critical Survey of Christian Attitudes Toward the World Religions*. Maryknoll, N.Y.: Orbis Books, 1985.

Kurzweil, Ray. *The Age of Spiritual Machines: When Computers Exceed Human Intelligence*. New York: Penguin Books, 1999.

————. *The Singularity Is Near: When Humans Transcend Biology.* New York: Penguin Books, 2006.

Lyons, J. A. *The Cosmic Christ in Origen and Teilhard de Chardin.* London: Oxford University Press, 1982.

Macquarrie, John. *Christology Revisited.* Harrisburg, Pa.: Trinity Press International, 1998.

Maloney, George M. *The Cosmic Christ: From Paul to Teilhard.* New York: Sheed & Ward, 1968.

Martelet, Gustave. *The Risen Christ and the Eucharistic World.* Translated by René Hague. New York: Seabury Press, 1976.

McFague, Sallie. *Life Abundant: Rethinking Theology and Economy for a Planet in Peril.* Minneapolis: Fortress Press, 2001.

Meilach, Michael D. *The Primacy of Christ in Doctrine and Life.* Chicago: Franciscan Herald Press, 1964.

————, ed. *There Shall Be One Christ.* New York: Franciscan Institute, 1968.

Menacherry, Cheriyan. *Christ: The Mystery in History, A Critical Study on the Christology of Raymond Panikkar.* Frankfurt am Mein: Peter Lang, 1996.

Merton, Thomas. *Conjectures of a Guilty Bystander.* New York: Doubleday Image, 1968.

————. *Contemplation in a World of Action.* New York: Image Books, 1973.

————. *Disputed Questions.* New York: Farrar, Straus & Cudahy, 1960.

————. *Introductions East and West.* Greensboro, N.C.: Unicorn Press, 1981.

————. *New Seeds of Contemplation.* New York: New Directions, 1961.

————. *No Man Is an Island.* New York: Harcourt Brace, 1955.

Miller, James B. *An Evolving Dialogue: Theological and Scientific Perspectives on Evolution.* Harrisburg, Pa.: Trinity Press International, 2001.

Miller, James, and Catherine Baker, eds. *The Evolution Dialogues: Science, Christianity, and the Quest for Understanding.* Washington, D.C.: American Association for the Advancement of Science, 2006.

Moltmann, Jürgen. *The Way of Jesus Christ: Christology in Messianic Dimensions.* Translated by Margaret Kohl. New York: HarperCollins, 1990.

Moravec, Hans. *Mind Children: The Future of Robot and Human Intelligence.* Cambridge, Mass.: Harvard University Press, 1990.

————. *Robot: Mere Machine to Transcendent Mind.* New York: Oxford University Press, 2000.

Noble, David F. *The Religion of Technology: The Divinity of Man and the Spirit of Invention.* New York: A. A. Knopf, 1997.

Ong, Walter. *Orality and Literacy: The Technologizing of the Word*. London: Methuen, 1982.

Osborne, Kenan B. *The Resurrection of Jesus: New Considerations for Its Theological Interpretation*. New York: Paulist Press, 1997.

Panikkar, Raimon. *Christophany: The Fullness of Man*. Translated by Alfred DiLascia. Maryknoll, N.Y.: Orbis Books, 2004.

———. "A Christophany for Our Time." *Theology Digest* 39, no. 1 (Spring 1992): 3-21.

———. *The Intra-Religious Dialogue*. New York: Paulist Press, 1978.

———. *Invisible Harmony*. Minneapolis: Fortress Press, 1995.

———. *Myth, Faith and Hermeneutics: Cross Cultural Studies*. New York: Paulist Press, 1980.

———. *The Unknown Christ of Hinduism*. Maryknoll, N.Y.: Orbis Books, 1981.

Peters, Ted. *Science, Theology, and Ethics*. Burlington, Vt.: Ashgate, 2003.

Polkinghorne, John. *Science and Theology*. Minneapolis: Fortress Press, 1998.

Prigogene, Ilya. *Order Out of Chaos*. Reissue edition. New York: Bantam Books, 1984.

Rahner, Karl. "Christology Within an Evolutionary View of the World." *Theological Investigations*. Volume 5. Translated by Karl-H Kruger. Baltimore: Helicon Press, 1966.

———. "Current Problems in Christology." *Theological Investigations*. Volume 1. Translated by Cornelius Ernst. Baltimore: Helicon Press, 1961.

———. "Natural Science and Reasonable Faith." *Theological Investigations*. Volume 21. Translated by Hugh M. Riley. New York: Crossroad, 1988.

———. "Science and Christian Faith." *Theological Investigations*. Volume 21. Translated by Hugh M. Riley. New York: Crossroad, 1988.

———. "The Two Basic Types of Christology." *Theological Investigations*. Volume 13. Translated by David Bourke. New York: Seabury Press, 1975.

Singh, Simon. *Big Bang: The Origin of the Universe*. New York: Harper, 2004.

Spink, Kathryn. *A Sense of the Sacred: A Biography of Bede Griffiths*. Maryknoll, N.Y.: Orbis Books, 1989.

Susin, Luiz Carlos, Andrés Torres Queiruga, and José María Vigil, eds. *Pluralist Theology: The Emerging Paradigm*. Concilium. Volume 1. London: SCM Press, 2007.

Swindells, John, ed. *A Human Search: Bede Griffiths Reflects on His Life: An Oral History*. Ligouri, Mo.: Triumph Books, 1997.

Teilhard de Chardin, Pierre. *Activation of Energy.* Translated by René Hague. New York: Harcourt Brace Jovanovich, 1970.

———. *Christianity and Evolution.* Translated by René Hague. New York: Harcourt Brace Jovanovich, 1971.

———. *The Divine Milieu: An Essay on the Interior Life.* Translated by William Collins. New York: Harper & Row, 1960.

———. *The Heart of Matter.* Translated by René Hague. New York: Harcourt Brace Jovanovich, 1978.

———. *Phenomenon of Man.* Translated by Bernard Wall. New York: Harper & Row, 1959.

———. *Science and Christ.* Translated by René Hague. New York: Harper & Row, 1968.

Thompson, William M. *Christ and Consciousness: Exploring Christ's Contribution to Human Consciousness.* New York: Paulist Press, 1977.

———. *Jesus, Lord and Savior: A Theopatic Christology and Soteriology.* New York: Paulist Press, 1980.

Tuckett, Christopher M. *Christology and the New Testament: Jesus and His Earliest Followers.* Louisville, Ky.: Westminster John Knox Press, 2001.

Toolan, David. *At Home in the Cosmos.* Maryknoll, N.Y.: Orbis Books, 2003.

Wessels, Cletus. *Jesus in the New Universe Story.* Maryknoll, N.Y.: Orbis Books, 2003.

Wheatley, Margaret. *Leadership and the New Science: Learning about Organization from an Orderly Universe.* San Francisco: Berrett-Koehler Publishers, 1992.

Wilber, Ken. *Sex, Ecology, Spirituality: The Spirit of Evolution.* Boston, Mass.: Shambala Publications, 2000.

———. *Up from Eden: A Transpersonal View of Human Evolution.* Wheaton, Ill.: Quest Books, 1996.

Wildiers, N. Max. *The Theologian and His Universe: Theology and Cosmology from the Middle Ages to the Present.* Translated by Paul Dunphy. New York: Seabury Press, 1982.

Worthing, Mark William. *God, Creation and Contemporary Physics.* Minneapolis: Fortress Press, 1996.

Wright, N. T. *Jesus and the Victory of God.* Vol. 2. *Christian Origins and the Question of God.* Minneapolis: Fortress Press, 1996.

———. *Simply Christian: Why Christianity Makes Sense.* New York: HarperCollins, 2006.

Wright, Robert. *Non Zero: The Logic of Human Destiny.* New York: Vintage, 2001.

Zagano, Phyllis. *Twentieth-Century Apostles: Contemporary Spirituality in Action.* Minneapolis: Liturgical Press, 1999.

INDEX